Staff Developers Manual

STATISTICS: A Key to Better Mathematics

The University of North Carolina
Mathematics and Science Education Network

Dale Seymour Publications®

Project Editor: Joan Gideon
Production Coordinator: Claire Flaherty
Design Manager: Jeff Kelly
Text and Cover Design: Don Taka

Published by Dale Seymour Publications®, an imprint of
the Alternative Publishing Group of Addison-Wesley Publishing Company.

The Teach-Stat materials were prepared with the support of National Science Foundation Grant No. TPE-9153779. Any opinions, findings, conclusions, or recommendations expressed in this publication are those of the authors and do not necessarily represent the views of the National Science Foundation. These materials shall be subject to a royalty-free, irrevocable, worldwide, nonexclusive license in the United States Government to reproduce, perform, translate, and otherwise use and to authorize others to use such materials for Government purposes.

Additional funding was provided by the North Carolina Department of Public Instruction and the North Carolina Statewide Systemic Initiative.

Order Number DS 21439
ISBN 1-57232-144-X

1 2 3 4 5 6 7 8 9 10-ML-99 98 97 96 95

This book is printed on recycled paper.

Co-Principal Investigators: Statewide Implementation

Susan N. Friel, UNC Mathematics and Science Education Network
Jeane M. Joyner, North Carolina Department of Public Instruction

Editors

Jane M. Gleason, North Carolina State University
Elizabeth M. Vesilind, University of North Carolina at Chapel Hill
Susan N. Friel, UNC Mathematics and Science Education Network
Jeane M. Joyner, North Carolina Department of Public Instruction

Co-Principal Investigators: Local Implementation

George W. Bright, University of North Carolina at Greensboro
Theresa E. Early, Appalachian State University
Dargan Frierson, Jr., University of North Carolina at Wilmington
Jane M. Gleason, North Carolina State University
M. Gail Jones, University of North Carolina at Chapel Hill
Robert N. Joyner, East Carolina University
Nicholas J. Norgaard, Western Carolina University
Mary Kim Prichard, University of North Carolina at Charlotte
Clifford W. Tremblay, Pembroke State University

Evaluation Team

Sarah B. Berenson, North Carolina State University
George W. Bright, University of North Carolina at Greensboro
Diane L. Frost, Curriculum and Instruction, Asheboro City Schools, Asheboro, North Carolina
Lynda Stone, University of North Carolina at Chapel Hill

Curriculum Consultants

Gwendolyn V. Clay, Meredith College
Gary D. Kader, Appalachian State University
Karan B. Smith, University of North Carolina at Wilmington

Editorial and Production Consultants

Stephanie J. Botsford, UNC Mathematics and Science Education Network
Stacy L. Otto, UNC Mathematics and Science Education Network

Administrative Support

Sherry B. Coble, Center for Mathematics and Science Education, University of North Carolina at Chapel Hill
Kathleen J. Gillespie, UNC Mathematics and Science Education Network
Jane W. Mitchell, UNC Mathematics and Science Education Network
Maurice A.Wingfield, Jr., Mathematics and Science Education Center, Fayetteville State University

Graduate Assistants

Anita H. Bowman, University of North Carolina at Greensboro
Lynne C. Gregorio, North Carolina State University
Sarah J. Pulley, East Carolina University
Julie C. Shouse, University of North Carolina at Wilmington
Kathryn L. Speckman, University of North Carolina at Greensboro
Kimberly I. Steere, University of North Carolina at Chapel Hill

Teach-Stat Advisory Board

Gloria B. Barrett, North Carolina School for Science and Mathematics
Randy L. Davis, Glaxo, Inc.
Denis T. DuBay, North Carolina Science and Mathematics Alliance
Diane L. Frost, Curriculum and Instruction, Asheboro City Schools, Asheboro, North Carolina
Manuel Keepler, North Carolina Central University, Department of Mathematics and Computer Science
L. Mike Perry, Mathematics Department, Appalachian State University
Sherron M. Pfeiffer, Equals, Hendersonville, North Carolina
Jerome Sacks, National Institution of Statistical Science
Michael J. Symons, Department of Biostatistics, University of North Carolina at Chapel Hill
John L. Wasik, Department of Statistics, North Carolina State University
Johnny F. Warrick, Robinson Elementary School, Gastonia, North Carolina

UNC Mathematics and Science Education Network Center Directors

Sarah B. Berenson, North Carolina State University
J. Ralph DeVane, Western Carolina University
Steven E. Dyche, Appalachian University
Leo Edwards, Jr., Fayetteville State University
Vallie W. Guthrie, North Carolina A & T State University
Katherine W. Hodgin, East Carolina University
Russell J. Rowlett, University of North Carolina at Chapel Hill
Josephine D. Wallace, University of North Carolina at Charlotte
Charles R. Ward, University of North Carolina at Wilmington

Contents

Introduction to the Staff Developers Manual

This part of the Teach-Stat program provides staff developers, such as university faculty and school administrators, with plans for a five-day Statistics Educators Institute. In these institutes, teachers who are already implementing Teach-Stat in their classrooms explore ways to help other teachers learn to use Teach-Stat. Upon completion of an institute, these teachers may serve as statistics educators able to plan and implement Teach-Stat workshops of varying lengths for teachers of grades 1 through 6. Leadership of Statistics Educators Institutes builds on and requires familiarity with *Teach-Stat for Teachers: Professional Development Manual*, the resource book for teacher workshops.

In the Statistics Educators Institute, several teachers work together to plan their first Teach-Stat workshop. We found this a successful first step in helping teachers make the transition from teaching children to teaching adults. A central component of the Statistics Educators Institute is supervised teaching by statistics educators during a regular two- or three-week Teach-Stat workshop. This manual is written as though two to eight statistics educators are sharing responsibility for planning and implementing a Teach-Stat workshop. The staff developer facilitates the statistics educators' planning and serves as a resource and coach when statistics educators present a Teach-Stat workshop for other teachers.

The Statistics Educators Institute is organized around nine modules, written for use by staff developers. Each module is a 2½- to 4-hour session. Modules may be combined in a number of ways, according to the experiences and needs of statistics educators. Each module begins with an overview of content, goals, and materials. Suggestions are made for facilitating the activities in the module, with key discussion questions and remarks italicized. Master copies of handouts follow each module.

In *Module 1, Setting Goals*, excerpts of teachers' talk about changes in teaching mathematics are used as vehicles for reviewing the *Professional Standards for Teaching Mathematics* (NCTM 1991). A panel discussion raises questions about ways to facilitate teacher change. Statistics educators practice introducing themselves and Teach-Stat, as if to a workshop audience.

Module 2, Planning for Teach-Stat Workshops, introduces a goal-based planning process in which Teach-Stat investigations are related to workshop objectives. Statistics educators draft outlines of goals, topics, and activities for a workshop they will facilitate. They also assess their consulting styles and, as a group, negotiate who will facilitate each activity during an upcoming workshop.

In *Module 3, Teachers as Learners,* statistics educators revise their workshop plans in light of research findings about adult learning and teacher change. Workshop presentation skills are related to needs of adult learners, and statistics educators set goals for improving their presentation skills.

Module 4, Issues in Teaching Statistics, focuses on self-assessment of statistics content knowledge, as statistics educators analyze concepts and procedures in Teach-Stat investigations and in local mathematics curriculum guides. From this assessment, the staff developer can plan for further instruction about statistics concepts to meet the needs of statistics educators.

In *Module 5, Planning the Details,* statistics educators prepare the final draft of the workshop agenda, write orders for materials and supplies, plan for formative and summative feedback, and assign a variety of workshop duties among themselves.

Module 6, Teaching Vignettes, presents vignettes of teaching situations, including teacher-student discourse about statistics concepts. Through discussion of the vignettes and role playing of classroom discourse, statistics educators enrich their pedagogical content knowledge as well as their own understanding of statistics concepts. Depending on statistics educators' needs, staff developers may decide to allot two sessions to this module.

In *Module 7, Presentation Skills,* statistics educators work with partners to rehearse and exchange constructive criticism. They plan strategies for responding to workshop participants' daily feedback and discuss ways of responding to any problems that arise.

Module 8, Preparing the Site, provides time for statistics educators to organize a workshop site, materials, and technology. Ideally this module occurs a day or two before a workshop begins.

Between *Modules 8* and *9,* statistics educators facilitate a Teach-Stat workshop they have planned for classroom teachers. This is a supervised internship for statistics educators, with the staff developer observing and serving as a resource. For example, in one successful project, a staff developer guided ten statistics educators who planned and facilitated a two-week summer workshop for 20 teachers.

Module 9, After the Workshop, is a follow up to the statistics educators' supervised leading of a Teach-Stat workshop. Participants' evaluations of the workshop are discussed. Statistics educators and the staff developer share ideas for promoting statistics educators as consultants in schools.

MODULE 1

Setting Goals

Overview

In *Activity 1, Warm-Up: "I used to . . . , but now I . . . ,"* statistics educators discuss how warm-up activities in workshops can set a tone of collegiality and establish facilitators' credibility.

Activity 2, Teacher Talk, provides a context in which to review the *Professional Standards for Teaching Mathematics* (NCTM 1991) and Teach-Stat goals. Statistics educators analyze testimony from teachers who have begun to implement Teach-Stat in their classrooms. Team building begins as statistics educators work in pairs to identify the Teaching Standards apparent in *Teacher Talk* handouts.

Activity 3, Introducing Teach-Stat in My Own Words, helps statistics educators decide how they will introduce themselves and Teach-Stat at the beginning of a workshop so workshop goals will be clear.

In *Activity 4, Panel Discussion: Teaching Teachers,* statistics educators with experience in leading workshops discuss questions about teaching teachers. The intent of this discussion is to raise issues that will be addressed in remaining modules of this manual.

The module concludes with *Activity 5, Organization of the Statistics Educators Institute,* which provides an overview and schedule of this program, including important clarification of roles.

Goals

Statistics educators will

- review the NCTM *Professional Standards for Teaching Mathematics* (1991) and the NCTM *Curriculum and Evaluation Standards for School Mathematics* (1989)
- discuss goals for a Teach-Stat workshop
- practice listening for teacher change in the direction of the NCTM Teaching Standards and Teach-Stat goals
- clarify purposes and agenda for this Statistics Educators Institute

Notes

Activity 1: 45 minutes

Activity 2: 60 minutes

Activity 3: 30 minutes

Activity 4: 45 minutes

Activity 5: 15 minutes

Curriculum and Evaluation Standards for School Mathematics. Reston, Virginia: National Council of Teachers of Mathematics, 1989.

Professional Standards for Teaching Mathematics. Reston, Virginia: National Council of Teachers of Mathematics, 1991.

A warm-up for groups who prefer more private sharing is called *Five Step.* The teachers stand in an open space in the room. Each teacher then takes five steps in any direction and selects as a partner another teacher he or she does not know well. The partners share their thoughts about a topic designated by the leader. For example, statistics educators might be asked to share their favorite Teach-Stat activity or how Teach-Stat has changed their teaching. After two minutes of sharing, the leader calls out "Five Step," and each teacher repeats the process with a new partner.

- clarify roles of the staff developer and statistics educators
- begin to build collegiality

Materials

- *NCTM Professional Standards for Teaching Mathematics* (Handout 1.1)
- *NCTM Mathematics Curriculum Standards* (Handout 1.2)
- *NCTM Mathematics Evaluation Standards* (Handout 1.3)
- *Teacher Talk* (Handout 1.4)
- *Possible Responses for Teacher Talk* (Handout 1.5, optional)
- *Teaching Teachers: Questions to Consider* (Handout 1.6)
- *Statistics Educators Institute: Modules* (add dates and times to create a schedule) (Handout 1.7)
- a journal for each statistics educator (optional)

Facilitating *Module 1*

It is important in *Module 1* for you, as staff developer, to take the role of facilitator rather than professor or deliverer of training. The first four activities establish statistics educators as sources of significant experience and knowledge and provide opportunities for them to develop skills in communicating this experience and knowledge. Throughout the Statistics Educators Institute, one focus is on the development of discourse about statistics. By beginning here in *Module 1* with statistics educators' discourse among themselves, you will be able to assess this discourse and determine what skills statistics educators may need to develop during the institute.

Activity 1 Warm-Up: "I used to . . . , but now I . . . "

Ask statistics educators to write a sentence in the form "I used to . . . , but now I . . . " The sentence should be about teaching. When ready, each person, beginning with the staff developer, tells his or her name, school, and grade level and then shares the sentence.

For example, you might say, "I used to avoid teaching box plots, but now I like them so much I dream about them."

What are some reasons for using warm-ups in workshops?

Workshop presenters establish their expertise and credibility through information shared during warm-ups. This eliminates the need to stand up and list one's credentials.

Warm-ups also set a tone of collegiality; they are not competitive or intimidating. Members of groups begin to build trust in each other. When participants are comfortable with each other, they are more likely to take the risks necessary for learning.

Activity 2 Teacher Talk

The discussion focus moves from personal experience with change to an analysis of other teachers' talk about change, helping statistics educators to relate the Standards and the Teaching Standards to these teachers' changes.

We've been sharing our own experiences with teaching and changes we've made in our own teaching. Now let's analyze how other teachers say they have changed. As we do that, we'll try to relate the Standards and the Teaching Standards to what each teacher reports. In this way we'll also be reviewing the Standards and the Teaching Standards, which, as statistics educators, we need to be able to explain to others.

Hand out *NCTM Professional Standards for Teaching Mathematics* (Handout 1.1), *NCTM Mathematics Curriculum Standards* (Handout 1.2), and *NCTM Mathematics Evaluation Standards* (Handout 1.3).

These handouts list standards for teaching, curriculum, and evaluation as recommended by the National Council of Teachers of Mathematics. They provide a vision of good mathematics teaching. We can review the Standards and the Teaching Standards by applying them within the context of teachers' talk.

Choose one of the four cases presented in *Teacher Talk* (Handout 1.4), and prepare it as a handout. In pairs or small groups, read and discuss the selected case. Discussion questions include:

Has this teacher changed in the direction of the **Professional Standards for Teaching Mathematics*****? How can you tell? Which Teaching Standards are reflected by this teacher? What other observations or inferences can you make from this teacher's talk?***

For examples of answers to these questions, see *Possible Responses for Teacher Talk* (Handout 1.5) for the selected case. This is an optional handout.

Variation: Warm-Up

An alternative warm-up is *Two Truths and a Lie.* Each person prepares three statements about his or her teaching or experience with statistics, two of which are true and one of which is false. As these are shared, the other participants must try to guess which statement is the lie. Example:

(1) *Changing the way I teach was easy for me.*
(2) *I've had four college-level statistics courses.*
(3) *Most of what I really know about statistics I've learned in Teach-Stat.*

Which statement is the lie?

Notes

The cases in *Teacher Talk* are elementary school teachers who permitted their reflections to be part of this manual. The names used here are pseudonyms.

Variation: Teacher Talk

Teacher Talk cases not used now may be read as part of a follow-up activity.

Notes

Teach-Stat goals are to help elementary-grades teachers (1) learn more about statistics and the process of statistical investigation, (2) use a statistical investigation model to enhance their teaching of mathematics, and (3) integrate statistics into science and social studies curricula.

A detailed description of the Teach-Stat project appears in the introduction to *Teach-Stat for Teachers: Professional Development Manual.*

Because statistics educators use Teach-Stat investigations in their own classrooms, their workshop presentations are highly credible to other teachers. Statistics educators are able to say, *When my students investigated this topic, these were their data, graphs, and interpretations.*

Variation: Panel Discussion

If statistics educators have had no experience in presenting to teachers, they may use the questions designed for the panel to interview a colleague who has led workshops.

Activity 3 Introducing Teach-Stat in My Own Words

One purpose of the **Teacher Talk** ***activity was to review the NCTM Standards and Teaching Standards. Why does a workshop presenter need to be familiar with these documents? Will you want to hand out these documents at a workshop?***

Discuss how workshop attendees expect expertise from leaders. When participants are required to attend a workshop, the workshop presenters may want to justify the value of the topic by referring to the Standards and Teaching Standards. Teach-Stat activities are intended to be models of the Teaching Standards.

Another reason for **Teacher Talk** ***is to show how powerful a professional development program like Teach-Stat can be in energizing teachers and creating significant changes in teaching.***

Ask statistics educators to write several words or phrases to describe how participation in a Teach-Stat workshop changed their teaching practices. Volunteers may share their reflections.

Now let's put all this into a workshop introduction. How will you tell other teachers what Teach-Stat is about? What will you say to introduce Teach-Stat at the beginning of a workshop?

Have statistics educators introduce themselves and Teach-Stat to a partner. Discuss any questions that emerge. Give statistics educators time to write ideas for workshop introductions in their notebooks.

Activity 4 Panel Discussion: Teaching Teachers

This discussion raises issues about being a workshop facilitator, issues that will be explored during the Statistics Educators Institute.

Before this session, choose about three participants who already have some experience in facilitating teacher workshops or in making presentations to other teachers. Give them *Teaching Teachers: Questions to Consider* (Handout 1.6), and ask them to prepare responses. Ask the panelists to save "horror stories" or scary comments about teaching one's peers until later in the institute. The goal here is to raise issues without creating anxiety for teachers who are less experienced.

Some of you are preparing to be teachers of teachers for the first time. Others have already facilitated teacher workshops. By questioning this panel, we can begin to share experiences.

The following questions for the panel are found on Handout 1.6:

- What surprised you the most when you started teaching teachers?
- How is teaching adults different from teaching children? How is it the same?
- What are similarities and differences between a teacher facilitating a workshop for other teachers and a district administrator or professor facilitating a workshop for teachers?
- As a workshop facilitator, what advantages do you have by also being a classroom teacher?
- What have been your best experiences in facilitating workshops?

During the discussion, assure an emphasis on positive experiences. Be ready to contribute your own positive experiences. Summarize by identifying topics raised in the discussion that will be addressed by this institute. This leads into *Activity 5*.

Activity 5 Organization of the Statistics Educators Institute

Hand out the chart *Statistics Educators Institute: Modules*, with dates and times added for your site (Handout 1.7).

This institute helps you prepare to organize and lead Teach-Stat workshops ranging from two hours to two or three weeks in length. The first eight modules will help you prepare for the workshop you will lead. The last module, which is recommended for use after you teach, will focus on evaluating your presentation efforts and looking ahead at issues with respect to your role as consultants.

This is a good time to discuss roles of everyone involved in preparing and presenting the workshop. If statistics educators already know you as a leader of earlier workshops, they need to view you now as coach and facilitator. They need to know that responsibility for planning and leading a workshop for other teachers will be theirs and that you will not lead investigations during that workshop. Statistics educators who do not understand your role as coach may be less active and may initiate fewer ideas than those who understand that *they* are responsible for planning and conducting the workshop.

Notes

Teach-Stat staff developers have found it helpful to *require* statistics educators to attend all institute and workshop sessions. Full participation promotes team building and equity in assignment of responsibilities. Regular attendance helps statistics educators to be well-prepared to present their workshop.

Variation: Modules

The sequence of the nine modules may be varied according to specific needs. For example, one staff developer extended *Module 6* by giving a brief presentation about statistics content during *every* module.

Another group used the following schedule to prepare a two-week July workshop: *Modules 1–5* on Saturdays and evenings during April and May, *Modules 6–8* during the week before the workshop, and *Module 9* over dinner after the workshop.

References

Curriculum and Evaluation Standards for School Mathematics. Reston, Virginia: National Council of Teachers of Mathematics, 1989.

Professional Standards for Teaching Mathematics. Reston, Virginia: National Council of Teachers of Mathematics, 1991.

NCTM Professional Standards for Teaching Mathematics

Standard 1: Worthwhile Mathematical Tasks

The teacher of mathematics should pose tasks that are based on

- sound and significant mathematics
- knowledge of students' understandings, interests, and experiences
- knowledge of the range of ways that diverse students learn mathematics

and that

- engage students' intellect
- develop students' mathematical understandings and skills
- stimulate students to make connections and to develop a coherent framework for mathematical ideas
- call for problem formulation, problem solving, and mathematical reasoning
- promote communication about mathematics
- represent mathematics as an ongoing human activity
- display sensitivity to, and draw on, students' diverse background experiences and dispositions
- promote the development of all students' dispositions to do mathematics

Standard 2: The Teacher's Role in Discourse

The teacher of mathematics should orchestrate discourse by

- posing questions and tasks that elicit, engage, and challenge each student's thinking
- listening carefully to students' ideas
- asking students to clarify and justify their ideas orally and in writing
- deciding what to pursue in depth from among the ideas that students bring up during a discussion
- deciding when and how to attach mathematical notation and language to students' ideas
- deciding when to provide information, when to clarify an issue, when to model, when to lead, and when to let a student struggle with difficulty
- monitoring students' participation in discussions and deciding when and how to encourage each student to participate

Standard 3: Students' Role in Discourse

The teacher of mathematics should promote classroom discourse in which students

- listen to, respond to, and question the teacher and one another
- use a variety of tools to reason, make connections, solve problems, and communicate
- initiate problems and questions
- make conjectures and present solutions
- explore examples and counterexamples to investigate a conjecture
- try to convince themselves and one another of the validity of particular representations, solutions, conjectures, and answers
- rely on mathematical evidence and argument to determine validity

Standard 4: Tools for Enhancing Discourse

The teacher of mathematics, in order to enhance discourse, should encourage and accept the use of

- computers, calculators, and other technology
- concrete materials used as models
- pictures, diagrams, tables, and graphs
- invented and conventional terms and symbols
- metaphors, analogies, and stories
- written hypotheses, explanations, and arguments
- oral presentations and dramatizations

Standard 5: Learning Environment

The teacher of mathematics should create a learning environment that fosters the development of each student's mathematical power by

- providing and structuring the time necessary to explore sound mathematics and grapple with significant ideas and problems
- using the physical space and materials in ways that facilitate students' learning of mathematics
- providing a context that encourages the development of mathematical skill and proficiency
- respecting and valuing students' ideas, ways of thinking, and mathematical dispositions

and by consistently expecting and encouraging students to

- work independently or collaboratively to make sense of mathematics
- take intellectual risks by raising questions and formulating conjectures
- display a sense of mathematical competence by validating and supporting ideas with mathematical argument

Standard 6: Analysis of Teaching and Learning

The teacher of mathematics should engage in ongoing analysis of teaching and learning by

- observing, listening to, and gathering information about students to assess what they are learning
- examining effects of the tasks, discourse, and learning environment on students' mathematical knowledge, skills, and dispositions

in order to

- ensure that every student is learning sound and significant mathematics and is developing a positive disposition toward mathematics
- challenge and extend students' ideas
- adapt or change activities while teaching
- make plans, both short- and long-range
- describe and comment on each student's learning to parents and administrators, as well as to the students themselves

Reference

Professional Standards for Teaching Mathematics. Reston, Virginia: National Council of Teachers of Mathematics, 1991, pp. 25–67.

NCTM Mathematics Curriculum Standards

Grades K–4

Mathematics as Problem Solving

Mathematics as Communication

Mathematics as Reasoning

Mathematical Connections

Estimation

Number Sense & Numeration

Concepts of Whole Number Operations

Whole Number Computation

Geometry & Spatial Sense

Measurement

Statistics & Probability

Fractions & Decimals

Patterns & Relationships

Grades 5–8

Mathematics as Problem Solving

Mathematics as Communication

Mathematics as Reasoning

Mathematical Connections

Number Sense & Number Relationships

Number Systems & Number Theory

Computation & Estimation

Patterns & Functions

Algebra

Statistics

Probability

Geometry

Measurement

Grades 9–12

Mathematics as Problem Solving

Mathematics as Communication

Mathematics as Reasoning

Mathematical Connections

Algebra

Functions

Geometry from a Synthetic Perspective

Geometry from an Algebraic Perspective

Trigonometry

Statistics

Probability

Discrete Mathematics

Conceptual Underpinnings of Calculus

Mathematical Structure

Reference

Guide to Standards: Presentation Materials. Reston, Virginia: National Council of Teachers of Mathematics, 1991.

NCTM Mathematics Evaluation Standards

Alignment

Methods and tasks for assessing students' learning aligned with the curriculum's goals, objectives, and content

Multiple Sources of Information

Decisions convey students' learning based on a convergence of information obtained from a variety of sources

Appropriate Assessment Methods and Use

Assessment methods and instruments selected on the basis of the type of information sought, the use to which the information will be put, and the developmental level of the students

Mathematical Power

Evidence about students' ability to solve problems, communicate mathematically, and reason and analyze; and the students' disposition toward mathematics

Problem Solving

Evidence that students are able to solve problems and generalize solutions

Communication

Evidence that students can express mathematical ideas by speaking, writing, demonstrating, and depicting the ideas visually

Reasoning

Evidence that students are able to use inductive and deductive reasoning

Mathematical Concepts

Evidence that students can label and define concepts and can compare, contrast, and translate one mode of conceptual representation to another

Mathematical Procedures

Evidence that students can reliably and efficiently execute procedures, recognize correct and incorrect procedures, generate new procedures, and extend or modify familiar procedures

Mathematical Disposition

Evidence that students are confident, show flexibility in thinking, show perseverance, and value the contribution of mathematics to other disciplines

Reference

Curriculum and Evaluation Standards for School Mathematics. Reston, Virginia: National Council of Teachers of Mathematics, 1989.
Guide to Standards: Presentation Materials. Reston, Virginia: National Council of Teachers of Mathematics, 1991.

Teacher Talk: Darcy

I've always been very, very weak in math. I've always felt like that was one of my worst subjects, because I didn't like math and never understood it. I didn't have someone to sit down with manipulatives and show me and let me understand what the concepts were. It was always I was taught "open the book and do these problems on this page." I taught 17 years and really didn't like math.

Learning to use manipulatives has changed my life as a teacher. I love manipulatives—and not just in math, but in all subjects. And I really became interested in and fell in love with math. And since then math is my favorite subject of all.

Three or four years ago I was using five books. We're down to three books now, and we've only used our social studies book a little bit. I use the math book once in a while for extra enrichment kinds of things, but a lot of my math is hands-on manipulatives, and my science is hands-on. We've been working on populations and planting plants, and they've charted how tall their plants grow and graphed. They've got to see that that's math, too, instead of just today, "turn to page 25 and do these ten problems on the page." I try to make it as real life as possible.

What I'm trying to do is change. I'm trying to become more flexible and let the students come up with things. I think I'm doing a better job. You know, before I would come up with all the questions and make sure I asked a certain list of questions. Teach-Stat has made me realize that sometimes I have to back off, and I have to let them go with it and see what they can do with it. You think a project is predictable, and this is the way it's going to be, and then it will come out something totally different in the end. I think Teach-Stat is helping me to be more flexible as a teacher. I'm beginning to see that I don't have to stay on everything every second of every day. I want to be able to just take something and go with it in statistical investigation. I'm working on that.

The only test scores I have are the standardized test scores we got last year, and I'll never be satisfied with my test scores. You think some of your children are going to do really well on something, and then they don't do as well. When they're taking the test and you walk around, you see some of the answers that they're putting down. Inside you say, "Oh, no! I know they know that, and we worked so hard on that." I just really don't think test scores are the best way to judge.

I use my camera and take pictures when the children are really doing something they're interested in. I put those into portfolios, because we're just starting portfolios. I'm really not as far along with the portfolios as I would like to be. They say change takes three to five years, and it's going to take me three to five years to get the portfolios to where I would like them to be. I wanted to start this year having student-led conferences and letting the students show the parents their portfolios, but I'm not quite far enough along right now. I'm hoping that at the end of school this year, I could let the students talk about their portfolios and show examples of the things they have learned this year.

People look at me funny when I say I taught 17 years and I hated math and then all of a sudden I got turned on to math. Math has come alive, and because of that, it has made all of the subjects come alive in my classroom.

Teacher Talk: Roland

The first year I taught, if it was 2:00, we definitely did science. And at 2:30 we did spelling, and at 3:00 you know, that sort of thing. And I was unhappy with that situation because I know that it just didn't all connect.

Considering the math strands, the seven basic things you need to teach in math, it was like the only thing I really knew about was the computation and numeration. And it hit me, "Oh, there are other things."

I've definitely grown, especially seeing how important not just the basic adding and subtracting, multiplying and dividing are. There are so many more critical-thinking and problem-solving skills that need to be introduced, and the children need experiences with these.

One area that we have struggled with this year is making sure everybody's needs are being met. And what we've done so far is given the kids lots of choices and different math projects, and we've let them take a project as far as they can go with it.

Every day each student has a math goal. And they record everything in a math journal; they write about math, they write about how they solve problems, they draw pictures about how they solve problems, and they're able to do this in a way that is best for them. The students choose when they do the math activities and how they do them. We say, for example, the next topic we are going to be studying is geology—rocks and minerals. And we say, what can you do to help us with this? So the kids might say, "Well, I can go out and collect soil samples" or "Could we call somebody and set up a field trip?" We give them things to find out about or they come up with a question they want to find out about, and they collect the data, and they organize it and explain it. We discuss it as a group.

If I used Teach-Stat in a more traditional way, I don't think kids would see connections between statistics and other subjects.

Teacher Talk: Marva

I've always liked math, but I've always liked teaching how to multiply and divide. But there's so much more now that we can do, so multiplying and dividing become secondary. We can go on and apply it, rather than trudge through all those multiplication tables all the time.

I am a traditional-type teacher who is still tied very much to her textbook. And I am still learning that if I don't cover what's done in the textbook, it's all right, as long as I'm working with the children on the process that I want them to achieve. But I feel I'm still struggling with the textbook and doing Teach-Stat. The first year I did this, I was trying to do too much. And as I go on with it, I've almost totally let go of the math book, so that I'm not worried about them seeing pages from the math book. But I am providing all the things that are in the curriculum that I need to cover. It doesn't necessarily have to be out of a math book. My biggest problem is with parents, because they expect to see a book or a worksheet come home every night. I'm trying to convince them that it's not necessary. When we had an open house, we did an investigation of the parents' favorite subject in school and the children's favorite subject. The majority of the parents said, "Oh, good. This sounds wonderful. I'm so excited." So I like it, as long as I don't get frustrated with trying to cover the state guidelines.

I'm seeing more problems with the discipline, because it's harder to keep them in control. They're so excited about a lot of the things that they do. And I like that excitement.

The new testing program is getting me to allow students to come up with as many ways as they can to answer a question, not just finding one answer. I think they are growing into that. For some it's very frustrating to have more than one solution, but they are becoming less afraid to try.

Teacher Talk: Rachael

We did some things with pumpkins in the fall. We were interested in the circumferences of pumpkins. And we did the circumference and talked about measuring. And then our question was, "Do we think that, based on the circumference of the pumpkin, the bigger ones would have more seeds?" And of course the kids were sure that the big ones had to have more seeds. So we actually counted. We cut open several pumpkins. Once we did that, we could see that the bigger pumpkins did not necessarily have more seeds than the smaller ones. And we talked about that quite a bit, and we graphed that. And a neat idea that came out: one girl said she had a belt of chain links and wanted to measure the pumpkins with chain links. They discovered that one of the pumpkins was as big around as the girl's waist. And we talked about how big around things are.

The positive aspects of being in Teach-Stat are understanding mathematical principles more and understanding how statistics fits into all areas of the curriculum and all that we do. We can look at statistics and some type of categorization of the data, and we can make it make more sense. Statistics really is the key to helping understand our world even better. And we can look at things we do in the classroom and base them on what they're doing out in the world and tie the world in through the vehicle of statistics. And I didn't understand that before. I didn't really think about that before we had Teach-Stat.

Possible Responses to Teacher Talk: Darcy

1. Darcy supports the use of a variety of instructional materials and models rather than simply teaching from the textbook. Physical materials give students something to think with and to discuss with others.
   - Standards for Teaching Mathematics: Standard 4 (tools for enhancing discourse)
   - Curriculum Standards: Standard 3 (the power of reasoning as a part of mathematics)

2. She makes interdisciplinary connections with populations and plants.
   - Standards for Teaching Mathematics: Standard 1 (worthwhile mathematical tasks)
   - Curriculum Standards: Standard 4 (connecting ideas within mathematics and between mathematics and other intellectual ideas)

3. She engages students in meaningful discourse rather than engaging in teacher-directed questioning.
   - Standards for Teaching Mathematics: Standard 2 (orchestrating discourse)
   - Curriculum Standards: Standard 2 (discussing mathematical ideas and making conjectures and convincing arguments)

4. She allows more flexibility in the direction a lesson takes and takes cues from students' interests and questions.
   - Standards for Teaching Mathematics: Standard 2 (orchestrating discourse by deciding what to pursue in depth from among the ideas students bring up in discussion)

5. She recognizes limitations of standardized testing.
   - Evaluation Standards: Standard 2 (multiple sources of information; appropriate assessment methods)

6. She uses portfolios.
   - Evaluation Standards: Standard 2 (multiple sources of information; appropriate assessment methods)

7. Students are involved in assessment.
   - Evaluation Standards: Standard 3 (appropriate assessment methods)
   - Darcy notes the frustration teachers often experience when required assessment methods do not align with the curriculum and instruction;

her comments also seem to reflect the importance still placed on standardized achievement scores.
- Evaluation Standards: Standard 1 (alignment)

8. Darcy shows a change in attitude toward mathematics.
   - Standards for the Professional Development of Teachers of Mathematics: Standard 5 (valuing mathematics)

Other comments about Darcy's talk

- She seems to realize that change is a process and not an event.
- She has noted how much she has changed in her attitude toward mathematics, but, at least in this excerpt, we cannot be sure of how much carry-over her attitude has with her students. She says, "I love manipulatives," but adds that her students have to "see that's math, too."

References

Curriculum and Evaluation Standards for School Mathematics. Reston, Virginia: National Council of Teachers of Mathematics, 1989.

Professional Standards for Teaching Mathematics. Reston, Virginia: National Council of Teachers of Mathematics, 1991.

Possible Responses to Teacher Talk: Roland

1. Roland makes interdisciplinary connections.
   - Standards for Teaching Mathematics: Standard 1 (worthwhile mathematical tasks)
   - Curriculum Standards: Standard 4 (connecting ideas within mathematics and between mathematics and other intellectual ideas)

2. He teaches from a broad mathematics curriculum.
   - Curriculum and Evaluation Standards: the 13 process and content areas identified for the K–4 curriculum (see Handout 1.2). (Note: Roland mentions "seven basic things," which may refer to the seven strands of the *North Carolina Standard Course of Study.*)

3. He values critical thinking and problem solving.
   - Curriculum Standards: Standard 3 (reasoning mathematically and having all students become problem solvers

4. He includes all students' needs and differentiates tasks for diverse needs.
   - Standards for Teaching Mathematics: Standard 1 (pose tasks that display sensitivity to, and draw on students' diverse background experiences and dispositions; mathematics for all students)

5. Roland's students use math journals.
   - Evaluation Standards: Standard 2 (multiple sources of information; appropriate assessment methods)

6. He poses questions to investigate a topic.
   - Standards for Teaching Mathematics: Standard 1 (tasks that call for problem formulation, problem solving, and mathematical reasoning)
   - Curriculum Standards: Standard 1 (problem solving as the focus of school mathematics)

References

Curriculum and Evaluation Standards for School Mathematics. Reston, Virginia: National Council of Teachers of Mathematics, 1989.

Professional Standards for Teaching Mathematics. Reston, Virginia: National Council of Teachers of Mathematics, 1991.

Possible Responses to Teacher Talk: Marva

1. Marva believes that mathematics is more than computation.
   - Curriculum and Evaluation Standards: The 13 process and content areas identified for the K–4 curriculum (see Handout 1.2)

2. Teaching the processes and concepts recommended in the curriculum is more important to Marva than covering the textbook. She supports the use of a variety of instructional materials.
   - Standards for Teaching Mathematics: Standard 4 (tools for enhancing discourse)

3. Marva realizes that changing the expectations of parents and that spreading the word to the public about goals of the Standards and the types of tasks they imply are important.
   - Standards for the Professional Development of Teachers of Mathematics: Standard 6 (participate in school, community, and political efforts to effect positive change in mathematics education)

4. Marva poses tasks that have more than one solution or answer; she promotes communication and student discourse as students try to convince classmates of the validity of their own solutions.
   - Standards for Teaching Mathematics: Standard 1 (tasks) and Standard 3 (discourse)

5. Marva's students take risks as they explore alternative solutions to questions.
   - Standards for Teaching Mathematics: Standard 5 (learning environment)

References

Curriculum and Evaluation Standards for School Mathematics. Reston, Virginia: National Council of Teachers of Mathematics, 1989.

Professional Standards for Teaching Mathematics. Reston, Virginia: National Council of Teachers of Mathematics, 1991.

Possible Responses to Teacher Talk: Rachael

1. Rachael helps students to redefine questions or pose new questions as they arise from an activity.
   - Standards for Teaching Mathematics: Standard 1 (pose tasks that call for problem formulation, problem solving, and mathematical reasoning) and Standard 5 (create an environment that encourages students to take intellectual risks by raising questions and formulating conjectures)

2. She understands that graphs are ways to communicate mathematically.
   - Curriculum Standards: Standard 2 (mathematics as communication)

3. She sees that the investigation about pumpkins is powerful. It has the potential to help children explore a variety of meaningful mathematics topics—circumference and standard versus nonstandard measures.
   - Standards for Teaching Mathematics: Standard 1 (pose tasks that are based on sound mathematics and that stimulate connections)

4. Participating in Teach-Stat gave Rachael the opportunity to further develop her own knowledge in mathematics. She came to value statistics as a tool for understanding the world.
   - Curriculum and Evaluation Standards: Goal 1 (valuing mathematics)
   - Standards for the Professional Development of Teachers of Mathematics: Standard 2 (knowing mathematics and school mathematics)

References

Curriculum and Evaluation Standards for School Mathematics. Reston, Virginia: National Council of Teachers of Mathematics, 1989.

Professional Standards for Teaching Mathematics. Reston, Virginia: National Council of Teachers of Mathematics, 1991.

Teaching Teachers: Questions to Consider

What surprised you the most when you started teaching teachers?

How is teaching adults different from teaching children? How is it the same?

What are similarities and differences between a teacher facilitating a workshop for other teachers and a district administrator or professor facilitating a workshop for teachers?

As a workshop facilitator, what advantages do you have by also being a classroom teacher?

What have been your best experiences in facilitating workshops?

Statistics Educators Institute: Modules

1	Setting Goals	Teaching Vignettes	6
2	Planning for Teach-Stat Workshops	Presentation Skills	7
3	Teachers as Learners	Preparing the Site	8
4	Issues in Teaching Statistics	After the Workshop	9
5	Planning the Details		

MODULE 2

Planning for Teach-Stat Workshops

Overview

This module helps statistics educators use a planning process in which the needs of workshop participants and the selected Teach-Stat objectives are matched with workshop activities. As statistics educators go through this planning process, they also review statistics content and begin to assess their own strengths and weaknesses related to the content.

In *Activity 1, Loved It/Hated It: What Makes an Effective Workshop?* statistics educators draw from their experience to describe a well-organized workshop. In *Activity 2, Planning a Conceptually Powerful Teach-Stat Workshop,* they create an agenda for the workshop that they will facilitate. *Activity 3, Evaluating Sample Workshop Outlines,* provides a review of criteria for effective workshop plans. In *Activity 4, Who Will Teach What?* and *Activity 5, Next Steps,* statistics educators decide who will assume responsibility for each aspect of the workshop.

Goals

Statistics educators will

- identify criteria of effective workshops
- review Teach-Stat concepts and activities
- use an understanding of participants' needs to set objectives for a Teach-Stat workshop
- match objectives with Teach-Stat activities
- write a preliminary agenda for a workshop
- assess their individual strengths and weaknesses in readiness to present workshops
- consider how to organize for teaching or, if more than one presenter will be conducting a workshop, how to assign teaching responsibilities

Notes

Activity 1: 20 minutes

Activity 2: 120 minutes

Activity 3: 30 minutes

Activity 4: 60 minutes

Activity 5: 20 minutes

Allow ample time for *Activities 2 and 4,* which contain the heart of the planning process.

Variation: Workshops of Different Lengths

Depending on participants' needs, *Module 2* may include planning hypothetical workshops of various lengths. If statistics educators are already committed to deliver a workshop, that immediate task can be the focus of this module, with references made to workshops of different lengths and purposes. In *Module 9* statistics educators have another opportunity to discuss workshops of different lengths.

Materials

- newsprint and markers for every 3 statistics educators (*Activity 1*)
- *Factors in Planning a Well-Organized Teach-Stat Workshop* (Handout 2.1)
- background information about possible or actual audience for whom a workshop is being planned, supplied by the staff developer (to accompany Handout 2.1)
- *Workshop Planning Template* (Handout 2.2)
- *Investigation Matrix* (Handout 2.3)
- *Planning for Teach-Stat Workshops* (Handout 2.4)
- *Criteria for Effective Workshops* (Handout 2.5)
- *Sample Plan for an After-School Workshop* (Handout 2.6)
- *Sample Plan for a Six-Hour Workshop* (Handout 2.7)
- *Ways to Assign Teaching Responsibilities and Partners* (Handout 2.8)
- *Consulting Style Inventory* (Handout 2.9)
- *Checklist for Planning Teach-Stat Workshops* (Handout 2.10)
- *Assignment* (Handout 2.11)
- local mathematics curriculum guides (*Activity 2*)
- *Teach-Stat for Teachers: Professional Development Manual* (*Activity 2*)

Facilitating *Module 2*

Staff developers help statistics educators take the lead in selecting objectives and activities for workshops. As statistics educators take on more responsible roles, try to resist when they say, "Just tell us which activities to do and we'll do them." At times you may need to slow down the planning to allow discussion about objectives. You can model questions that statistics educators need to ask as they plan (examples of helpful questions are given in this module). As statistics educators think about who will teach what, you should encourage self-assessment of content knowledge and presentation styles. Self-assessment will help statistics educators to form effective, well-rounded teaching teams.

Notes

Statistics educators take the lead in planning, while the staff developer serves as coach, questioner, and facilitator. Adoption of these roles during *Module 2* is essential to a successful workshop later.

Activity 1 Loved It/Hated It: What Makes an Effective Workshop?

One reason preparing for a workshop takes so much time and energy is that most consultants and presenters want participants to feel that workshops are good experiences for teachers. What exactly constitutes a good workshop experience?

Statistics educators form groups of three or four people. In their groups they list things they have personally loved or hated about workshops they have attended. Lists may be made on large sheets of newsprint and results shared after five minutes. Allow time to share stories.

In workshops that you remember, were all participants equally involved?

Sometimes a certain part of a workshop may really engage one teacher and not interest another teacher. Presenters need to accept that sometimes they cannot please everyone.

Which of the items on your lists were controlled by the workshop presenter?

Some features that make a workshop effective or ineffective are controlled by the presenter, such as the workshop organization, knowledge of subject matter, and effective communication skills. Some features may be more difficult to control, such as the facility itself or the starting and ending times.

What do you think is meant when a workshop is described as "well-organized"?

If organization is not specifically mentioned on a list, ask, "What about workshop organization?" Teachers often state that they like workshops to be "well-organized." If this phrase is offered, probe further by asking, "What do you mean that the workshop was well-organized?"

Summarize the *Loved It/Hated It* activity. Point out that two essential components of effective workshops are (1) knowledge of subject matter and (2) organization. The remainder of this module will focus on these topics.

Notes

Examples of *Loved It!*

enthusiastic leader
knew her topic
useful for my classroom
related to our curriculum
hands-on
really learned

Examples of *Hated It!*

too much wasted time
no examples
not enough interaction
disorganized

Activity 2 Planning a Conceptually Powerful Teach-Stat Workshop

Local curriculum guides are additional resources for selecting topics to include in a workshop.

As workshop topics are determined, the staff developer also helps statistics educators see that in a well-organized workshop statistics concepts are introduced in a logical sequence and activities relate conceptually to one another.

Sometimes when participants say that a workshop was well-organized, they are saying they felt that all the pieces of the workshop fit together. While a workshop full of one dazzling activity after another can be stimulating, unless certain considerations have been made in planning, it may do little to develop understanding of the topic or subject matter. A conceptually powerful workshop that also includes great activities can be achieved through careful planning.

Distribute *Factors in Planning a Well-Organized Teach-Stat Workshop* (Handout 2.1), *Workshop Planning Template* (Handout 2.2), *Investigation Matrix* (Handout 2.3), and *Planning for Teach-Stat Workshops* (Handout 2.4).

Discuss items 1 through 5 on Handout 2.1 and add details pertinent to your workshop. Handout 2.1 includes directions for using Handouts 2.2, 2.3, and 2.4. This planning activity may take two to three hours.

Summarize *Activity 2* by reviewing points on *Criteria for Effective Workshops* (Handout 2.5).

Activity 3 Evaluating Sample Workshop Outlines

Activity 3 may be used now or saved for *Module 9* when future workshops of varying lengths are discussed.

Distribute *Sample Plan for an After-School Workshop* (Handout 2.6) and *Sample Plan for a Six-Hour Workshop* (Handout 2.7). Discuss these plans in light of information given in Handout 2.5.

What do you like about this workshop outline?

How might you improve this workshop outline?

Activity 4 Who Will Teach What?

It is often desirable for statistics educators to co-present or team-teach a Teach-Stat workshop. In *Activity 2*, initial selections of objectives and investigations for the workshop were made. Now statistics educators will want to determine their specific teaching responsibilities.

Determining *who will teach what* is a process that must be done by the statistics educators themselves. You, the staff developer, will need to make sure these decisions are made fairly and equitably. For example, you will want to discourage the practice of assigning what may be viewed as all of the "easy parts" to one teacher and all of the "hard parts" to another. One option is to ask each statistics educator to take a familiar topic to teach as well as a topic that is less familiar. In this way statistics educators can expand their repertoire of "easy parts" to teach.

As a group we now have selected our Teach-Stat objectives and related investigations. The next step in our planning is to decide who will be responsible for presenting each part of the workshop. How could we do this?

On newsprint make a list of all suggestions offered, pointing out the features of each method suggested. A summary of methods for assigning responsibilities and features of those methods is provided in *Ways to Assign Teaching Responsibilities and Partners (*Handout 2.8).

Introduce the *Consulting Style Inventory* (Handout 2.9) as a tool for helping to assign workshop responsibilities.

Each method of assigning teaching responsibilities has advantages and disadvantages. Because one of the goals of this Statistics Educators Institute is to help you develop skills as a consultant, we will take an analytical approach to this task.

Basketball coaches begin successful seasons by recruiting top players, but they also watch their teams practice so that they can make good, informed position assignments.

We are starting with a group of Teach-Stat "All-Americans." We, too, want to make good teaching teams and assignments for our workshops. One way to help us make good teaching teams is to conduct a self-assessment.

Hand out the *Consulting Style Inventory*. Ask statistics educators to complete the form individually.

The purpose of this inventory is to help statistics educators decide how they want to pair up to teach during the workshop. Note that the *Investigation Matrix* (Handout 2.3) may help in responding to the first item on the inventory.

After statistics educators complete the *Consulting Style Inventory*, results may be shared. Remind statistics educators that the goal is to assess a team's strengths by first identifying strengths among its individual members.

Now that we have information about our individual strengths, we can address another dimension of finding good teaching partners.

Notes

When a workshop has multiple presenters, sometimes it is difficult to separate a discussion about what to teach and a discussion about who will teach it. Once a presenter either chooses or is assigned a topic, he or she may be reluctant to let go of the topic if the workshop agenda has to be changed.

Experienced Teach-Stat staff developers strongly recommend that deciding who will present each topic be delayed until the workshop agenda is finalized.

Variation: Inventory

Even if the workshop will be presented by only one statistics educator, the *Consulting Style Inventory* may be helpful for self-analysis.

Some people say that the best partners are those who are alike. Others say that diversity is a strength in itself.

Do you think we should form teams of presenters who are alike or different? What are advantages and disadvantages of each arrangement?

Following a discussion about how to form teams, return to *Planning for Teach-Stat Workshops* (Handout 2.4), where workshop topics have been identified.

Now we can look at the topics and investigations we have chosen for our workshop and decide, as a group, who should present each one. We will base these decisions upon what we now know about each other and on our beliefs about effective partnerships.

Below are key points to consider in forming workshop partnerships and assigning workshop topics.

- Be sure the group can justify each teaching assignment.
- Make sure each presenter has the opportunity to teach "something old and something new"—something that the presenter is familiar with or has taught to children and something that will make the presenter grow professionally.
- Try to give statistics educators opportunities to present with a variety of colleagues, so that when they prepare to conduct future workshops, they will have better ideas about who might make good partners for them.

Activity 5 Next Steps

We have built the framework for the workshop. Now that we have decided who will teach what, we need to fill in other details.

Hand out *Checklist for Planning Teach-Stat Workshops* (Handout 2.10).

Assist statistics educators as they add necessary items to this basic checklist, check off items already assigned, and make assignments for other items. Set deadlines as necessary. On your own copy of the checklist, record who is responsible for each item. At the next session you may want to distribute copies of this list of responsibilities.

Distribute *Assignment* (Handout 2.11).

Complete the assignment described in Handout 2.11 prior to the next workshop-planning meeting. This will help you prepare for your parts of the workshop.

Notes

Make sure there is equitable distribution of teaching and nonteaching assignments. Every statistics educator should have teaching assignments.

Factors in Planning a Well-Organized Teach-Stat Workshop

1. *Audience.* A first step in planning effective instruction is to learn about your audience. What types of teachers will be in the workshop? What grades or subjects do they teach? What sorts of backgrounds in statistics do they have?

2. *Time frame.* Confirm the length of time provided for your workshop. The *Workshop Planning Template* (Handout 2.2) may help organize workshop days.

3. *Goals.* Next define your goals. Keeping the needs of your audience in mind, what are the goals of your Teach-Stat workshop? Make a specific list of what you want the participants to know and to be able to do at the end of the workshop. Try to start each statement on the list with the stem "The teachers will be able to . . . ," which focuses on *objectives* rather than on *activities.*

4. *Objectives.* Now clarify your initial list of workshop objectives by using the *Investigation Matrix* (Handout 2.3). This matrix provides cross-references of major topics in statistics with the investigations in *Teach-Stat for Teachers: Professional Development Manual.* Review the topics in the matrix. Which of these topics do you want to include in your list of workshop objectives?

   As you review Teach-Stat topics for selection or elimination, it may be helpful to ask yourself the following questions:

   a. Is this topic essential for my audience, given our time frame, or is it one that would be nice if we had more time?

   b. Why is this topic appropriate for this audience?

   c. If this topic is omitted, will it result in a gap in the participants' conceptions related to statistics?

   A local curriculum guide may assist you in deciding which statistics topics do not relate to the teaching needs of your participants.

5. After using the *Investigation Matrix* to think about workshop topics, revise your original list of topics. It helps further planning to list activity names next to your objectives using *Planning for Teach-Stat Workshops* (Handout 2.4).

   Some consultants believe it is a good practice to ask workshop participants either to express their perceived needs or to have some actual input into workshop objectives and topics. Ask for their ideas early in your workshop, perhaps when you communicate your goals and objectives to the group.

Workshop Planning Template

	Day 1	Day 2	Day 3	Day 4	Day 5
A.M.					
P.M.					

	Day 6	Day 7	Day 8	Day 9	Day 10
A.M.					
P.M.					

HANDOUT 2.3

INVESTIGATION MATRIX

Topic	1 About Us	2 Sorting People: Who Fits My Rule?	3 Yekttis	4 Sorting Things!	5 Restructuring Mathematics	6 Shape of the Data: Using Line Plots	7 Shape of the Data: Line Plots/Bar Graphs	8 Giant Steps, Baby Steps	9 Typical Foot Length of Our Group	10 Children and Measurement	11 Accuracy in Measurement	12 How Close Can You Get to a Pigeon?
Posing questions	●						●		●			
Methodology	●											●
Measurement								●	●	●	●	●
Probability												
Sampling												
Types of data/scales								●	●		●	
Descriptive statistics												●
Median												
Mean												
Mode						●						
Range						●						
Representations	●											●
Bar charts												
Bar graphs		●					●					
Box plots												
Histograms												
Line plots						●	●		●			
Venn diagrams		●	●	●								
Scatter plots												
Stem-and-leaf plots												
Shape of the data						●						●
Sorting/classifying		●	●	●								
Describe/summarize	●					●						●
Compare/contrast	●					●						
Generalize	●					●						●
Identify associations												

HANDOUT 2.3

INVESTIGATION MATRIX

Topic	13 Family Size	14 Median: More Than Just the Middle	15 Types of Data: A Minilecture	16 About Us Revisited	17 Shape of the Data: Using Stem Plots	18 Shape of the Data: Stem Plots/Histograms	19 Curriculum Overview	20 Shape of the Data: Problem-Solving Sheet	21 Graphing: A Minilecture	22 Raisins Revisited	23 How Do We Grow?	24 Raisins Revisited Revisited
Posing questions	●											
Methodology												
Measurement												
Probability												
Sampling										●		
Types of data/scales			●									
Descriptive statistics										●		
Median		●		●							●	
Mean												
Mode				●								
Range		●		●								
Representations										●		
Bar charts	●								●			
Bar graphs	●							●				
Box plots											●	●
Histograms						●		●				
Line plots					●			●				●
Venn diagrams												
Scatter plots												
Stem-and-leaf plots					●	●		●				
Shape of the data					●					●		
Sorting/classifying												
Describe/summarize	●	●				●				●	●	
Compare/contrast		●								●	●	
Generalize	●					●				●		
Identify associations												

HANDOUT 2.3

INVESTIGATION MATRIX

Topic	25 Graphing Data Using Computers	26 Family Size Revisited	27 Cats	28 Technology	29 Building the "Rule" for Finding the Mean	30 Means in the News	31 Comparing Sets of Cereal Data	32 Student's Understanding of the Mean	33 Classroom Assessment of Mathematics	34 Linking Probability and Statistics	35 True-False Test	36 Removing Markers from a Number Line
Posing questions												
Methodology												
Measurement												
Probability										●	●	●
Sampling			●									
Types of data/scales			●									
Descriptive statistics			●									
Median							●					
Mean		●			●	●	●	●				
Mode												
Range												
Representations			●									
Bar charts												
Bar graphs												
Box plots							●					
Histograms												
Line plots							●					
Venn diagrams												
Scatter plots												
Stem-and-leaf plots												
Shape of the data			●			●						
Sorting/classifying				●								
Describe/summarize			●									
Compare/contrast			●				●					
Generalize			●				●					
Identify associations												

HANDOUT 2.3

INVESTIGATION MATRIX

Topic	37 What Are the Odds?	38 Fair Games	39 What's in the Bag?	40 Choosing Samples	41 How Tall Are You?	42 Are You a Square?	43 From Footprint to Stature	44 Cats Revisited				
Posing questions												
Methodology												
Measurement												
Probability	●	●	●	●								
Sampling												
Types of data/scales					●	●	●	●				
Descriptive statistics				●	●	●	●	●				
Median												
Mean												
Mode												
Range												
Representations												
Bar charts												
Bar graphs												
Box plots												
Histograms												
Line plots												
Venn diagrams												
Scatter plots					●	●	●	●				
Stem-and-leaf plots												
Shape of the data												
Sorting/classifying												
Describe/summarize				●	●	●	●	●				
Compare/contrast				●								
Generalize				●								
Identify associations					●	●	●	●				

Planning for Teach-Stat Workshops

Date______ Starting Time______ Ending Time______

Objectives/ Topics	Investigation/ Activity	Materials	Presenter	Time	Comments

Criteria for Effective Workshops

1. A well-organized workshop gives participants the feeling that what they have experienced is more than just an accumulation of neat, but perhaps unrelated, activities. As you plan, build in time to help participants make connections among activities, and from activities to objectives and goals.

2. Make sure the sequence of activities is developmentally sensible. For example, you would not want to do an activity on box plots before you do one on line plots. The table of contents in *Teach-Stat for Teachers: Professional Development Manual* may be helpful in determining presentation sequences.

3. The way the workshop starts is very important. Workshops that engage participants from the very beginning, perhaps with an activity, seem to be viewed as "better" by participants than those that start off with a lecture.

4. What is the purpose of your opening activity? Review *Module 1* for pointers about warm-ups.

5. A variety of levels of participant involvement is desirable. Workshops in which participants are continually involved in physical activities are just as unpopular as lecture-based workshops in which participants are always seated. Classify activities you have selected for your workshop as *active* or *sedentary*, and see if you are satisfied with the variety. For example, will you want a quiet activity right after a meal?

6. Another dimension of objectives and activities that needs to be considered is content level. For example, if the content of the activity is particularly challenging, participants may appreciate sessions of shorter duration with scheduled breaks. These breaks could be for reflection or just for fun.

7. The way the workshop day ends is important. Participants appreciate it when presenters provide some closure to the day with time to reflect upon the content of the workshop.

8. Provide time at the end of each workshop session for questions and comments. Writing in journals or on comment cards may be less threatening to participants than asking questions orally in the large group. On comment cards, participants anonymously give their reactions to the session and ask questions. This method gives the presenter feedback on participants' needs. But beware! Comment cards may, at least initially, contain a number of *basic needs* comments ranging from the temperature of the room to the refreshments provided. Like all forms of evaluation, comment cards require that the

recipient is ready to accept negative as well as positive feedback. By using comment cards, you are saying that you care about participants' needs, whatever they are, and that you are willing to act upon them. Ask for comments only if you plan to act on them, and allow time in your agenda to respond to comment cards at the next session.

9. Allow sufficient time at the conclusion of your workshop for overall reflection and that necessary evil, paperwork, but try not to make the very last thing you do with participants the paperwork. Ending with a positive feeling about statistics is just as important as a good beginning!

Sample Plan for an After-School Workshop

This workshop will take two to three hours.

I. ***Statistical Awareness***

Show magazine and newspaper examples

Share examples of statistics in everyday life

II. ***What Is Statistics?***

Define statistics

Introduce PCAI model

III. ***Overview of Curriculum***

List topics and activities that teachers presently use in their classrooms

Show objectives from state and local curriculum guides

Talk about national and local curriculum changes

IV. ***Statistical Investigations Using the PCAI Model and Types of Data Representations***

Sorting People: Who Fits My Rule?

Shape of the Data: Using Line Plots ■ Raisins

Shape of the Data: Using Stem Plots ■ Breath Holding

Building the "Rule" for Finding the Mean

Shape of the Data: Using Box Plots ■ Raisins Revisited

(Activities from *Teach-Stat for Teachers: Professional Development Manual)*

Sample Plan for a Six-Hour Workshop

8:45 Warm-up; posing questions and collecting data

- *What time did you go to sleep last night?* (Have participants sign in to the workshop with this information. Then try to make some generalizations.)

9:00 Overview

- the Teach-Stat project
- statistics and the process of statistical investigation—PCAI
- the need for change
- local curriculum guide (a quick review)

9:30 Sorting and classifying to analyze data

- *Guess My Rule*
- *Yekttis*

10:30 Break

10:45 Collecting data; representing data to analyze data

- bar charts and graphs
- *Family Size*

11:15 Measures of center

- overview of NAEP results
- *Building the "Rule" for Finding the Mean*

12:00 Lunch and sharing of teaching ideas

12:45 Collecting and representing data to analyze; line and stem plots

- *Breath Holding*

1:15 Pulling it all together

- *Raisins Revisited*

2:00 Distribution of materials (such as *Used Numbers* books)
Evaluation/comment cards

2:30 ¡Adios!

(Activities from *Teach-Stat for Teachers: Professional Development Manual*)

Ways to Assign Teaching Responsibilities and Partners

Method	*Features*
■ Dictate assignments and teaching partners	■ Not democratic ■ Efficient
■ Assign topics and partners randomly	■ Does not account for presenters' unique experiences, strengths, and personalities ■ Makes everyone learn to teach a topic he or she might eventually teach in a workshop
■ Ask for volunteers for assignments and partners	■ More assertive presenters may get their choices, less assertive may not ■ Takes advantage of presenters' unique interests and experiences ■ May be too many volunteers for some activities, too few for others ■ Teaching load may be inequitable ■ Some presenters may not have a teaching partner in mind ■ Teams made of friends may or may not have compatible grade-level experience ■ Partners who choose each other may cooperate better

Consulting Style Inventory

Part I: Take a few minutes to assess yourself in the following areas.

	none or low	some or medium	a lot or high
1. Credibility as a presenter with			
primary-grade teachers	0	1	2
upper elementary-grade teachers	0	1	2
middle-grade teachers	0	1	2
administrators	0	1	2
2. Experience presenting to adults	0	1	2
3. Experience with technology	0	1	2
4. Ability to attend to details; organizational skills	0	1	2
5. Ability to see the big picture	0	1	2
6. Sensitivity to the needs and feelings of others	0	1	2
7. Tendency to want to be in charge	0	1	2
8. Sense of fun, humor	0	1	2

Part II: Write answers to the following questions.

9. Do you have a dynamic, high-energy presentation style or a low-key style?

10. Are you a last-minute planner or a person who plans far in advance?

11. How do you react to suggestions from others?

12. Which Teach-Stat topics and activities do you feel most comfortable teaching right now?

13. Which Teach-Stat topics and activities do you feel least comfortable teaching right now?

14. What other considerations are important to you in choosing topics to teach?

15. What considerations are important to you in choosing a teaching partner?

Checklist for Planning Teach-Stat Workshops

___ Meet with person requesting workshop to discuss expectations and compensation.

___ Reserve a location, or make sure this has been done.

___ Select and meet with presenters.

___ Make a tentative timeline for preparations; schedule planning meetings.

___ Select workshop objectives.

___ Plan the workshop schedule to meet objectives and audience.

___ Determine presentation partners, if applicable.

___ Plan and prepare workshop activities.

___ Determine materials needed.

___ Gather materials, and photocopy the receipts.

___ Order books and other items needing purchase orders.

___ Arrange for offering certification renewal or graduate credit.

___ Publicize workshop, and have teachers sign up or apply. This may be done by the person requesting the workshop.

___ Create a database of participants; send acceptance letters. This may be done by the person requesting the workshop.

___ Assign participants to groups.

___ Reserve audiovisual equipment, computers, and software.

___ Review workshop agenda in light of participants' experiences and needs.

___ Prepare participants' packets, name tags, and handouts.

___ Remind participants about the workshop; send maps, parking information, and a list of any materials to bring.

___ Walk through the plans with partners; practice presentations and explanations; make transparencies.

___ Make sure the technology works—hardware and software.

___ Confirm audiovisual equipment reservations and location reservation; remind principals, janitors, or other building personnel; confirm access to the building and obtain any needed keys.

___ Purchase or arrange for refreshments.

___ Arrange the room the day before the workshop, if possible; post signs on outside and inside doors.

___ Photocopy attendance rosters and certificates for credit.

___ Pack materials the day before the workshop.

Assignment

1. Review the topics and activities that you will teach.

2. By a mutually agreed-upon date, write a list of supplies you will need, and give this list to whomever is responsible for ordering supplies.

3. If you have not already taught your topics to students, you may wish to do so. Remember to save examples of students' work, which you may use in the Teach-Stat workshop.

4. Begin to review and extend your knowledge of statistics. The following books may be helpful:

Graham, A. *Statistical Investigations in the Secondary School.* New York: Cambridge University Press, 1987.

Landwehr, J., and A. Watkins. *Exploring Data.* Rev. ed. Palo Alto, California: Dale Seymour Publications, 1995.

Moore, D. S. *Statistics.* 3d ed. New York: W. H. Freeman, 1991.

Newman, C., T. Obremski, and R. Scheaffer. *Exploring Probability.* Quantitative Literacy Series. Palo Alto, California: Dale Seymour Publications, 1987.

University of North Carolina Mathematics and Science Education Network. *Teach-Stat for Teachers: Professional Development Manual.* Palo Alto, California: Dale Seymour Publications, 1996.

Used Numbers: Real Data in the Classroom. Palo Alto, California: Dale Seymour Publications, 1989–1992.

MODULE 3

Teachers as Learners

Overview

Beliefs about teacher change that guide Teach-Stat workshops are consistent with a constructivist theory of learning. Constructivism maintains that learners bring prior knowledge and experiences to their highly individualized construction of concepts. This belief about learning suggests that teacher education is not simply demonstration of classroom activities. Teacher education helps teachers to focus on learners' conceptions and to make instruction responsive to these conceptions. As teachers begin to focus more closely on children's thinking, teachers may then begin to change how they use materials and methods (Schifter and Fosnot 1993).

This module is intended to help statistics educators understand why workshop participants should be actively engaged in reflecting on their own learning as they *do* statistics, why participants need to investigate their *own* questions and analyze their *own* data, why participants' apparent mistakes are opportunities for learning, and why leaders of Teach-Stat workshops are *coaches* rather than lecturers.

In *Activity 1, Cooperative Learning Jigsaw,* statistics educators study and share ideas about teachers as learners. Beliefs about learning discussed in *Activity 1* have significant implications for statistics educators' roles as workshop presenters.

In *Activity 2, Planning and Presenting for Adult Learners,* statistics educators evaluate their workshop plans in light of what has been learned about adult learners, teacher change, and team building.

During *Activity 3, Team Planning Time,* partners or teams revise their workshop plans, using criteria discussed in *Activities 1* and *2*. This is an opportunity for you, the staff developer, to make sure that workshop plans focus on active engagement of participants and not on *talking at* participants.

Activity 4, Reflection on One's Own Presentation Style, provides self-assessment of strengths and weaknesses in statistics educators' workshop presentation skills, along with specific planning for rehearsals.

Notes

Activity 1: 45 minutes

Activity 2: 30 minutes

Activity 3: 30 minutes

Activity 4: 30 minutes

Resources about constructivism:

Brooks, J. G., and M. G. Brooks. *In Search of Understanding: The Case for Constructivist Classrooms.* Alexandria, Virginia: Association for Supervision and Curriculum Development, 1993.

Confrey, J. "What Constructivism Implies for Teaching." In *Constructivist Views on the Teaching and Learning of Mathematics,* edited by R. B. Davis, C. A. Maher, and N. Noddings. Reston, Virginia: National Council of Teachers of Mathematics, 1990.

Schifter, D., and C. T. Fosnot. *Reconstructing Mathematics Education.* New York: Teachers College Press, 1993.

Goals

Statistics educators will

- identify needs of adult learners
- apply their understanding of adult learners, teacher change, and team building to the planning of Teach-Stat workshops
- identify workshop presentation skills that support adult learning and teacher change
- reflect on their own presentation skills—strengths and areas for improvement—and plan for improvement

Materials

Resources about presentation skills:

Bowers, P. S. "The Nuts and Bolts of Planning Workshops for Teachers." *Science Education* 3, no. 6 (1994): 27–31.

Sharp, P. A. *Sharing Your Good Ideas: A Workshop Facilitator's Handbook.* Portsmouth, New Hampshire: Heinemann, 1993.

Materials

- *When Teachers Learn* (Handout 3.1)
- *Expressions of Change* (Handout 3.2)
- *Team Building* (Handout 3.3)
- *Planning for Adult Learners* (Handout 3.4)
- the statistics educators' workshop plans (from *Module 2)*
- *My Presentation Skills* (Handout 3.5; if statistics educators have journals, they may write in those and do not need this handout)
- resource books about workshop presentation skills (optional)

Facilitating *Module 3*

Throughout *Module 3* you will find opportunities to help statistics educators apply findings from research and theory to their own workshop planning. Any tendency for statistics educators to talk too much at workshop participants or to not allow adequate time for participants to experience the statistical investigation process can be addressed by your referring to the readings in *Activity 1, Cooperative Learning Jigsaw.* Constructivist views of teaching and learning may be presented or reviewed in the context of discussing advantages and disadvantages of certain teaching strategies—lecturing, letting teachers do investigations, using small groups, asking questions. Teach-Stat staff developers report that statistics educators are usually skillful in direct instruction but that many need practice in teaching by asking questions. These topics often emerge again during discussions at the end of each workshop day, after statistics educators have observed each other's teaching and participants' learning.

Variation: Jigsaw

For a group of fewer than six statistics educators, copies of the three jigsaw articles may be distributed to everyone for reading and discussion.

Activity 1 Cooperative Learning Jigsaw

Up to now we have focused on statistics content of a Teach-Stat workshop. Now the focus shifts to what is known about how adults learn, how teachers change, the importance of team building, and how we can use this information to enhance workshops.

Introduce the activity.

We will use a jigsaw activity to help each other reflect on adult learning, how teachers change, and team building. Then we will apply these ideas to our workshop plans. This can help us to make our presentation methods appropriate for adult learners.

This jigsaw requires six or more participants and Handouts 3.1, 3.2, and 3.3. Directions for organizing the activity follow.

1. Assign statistics educators as equally as possible to three home-base teams, and ask those teams to gather.

2. Ask each team to assign its members as evenly as possible to three expert groups—one about adult learning, one about teacher change, and one about team building. Any extra members should be assigned to the adult learning topic. Each participant is thus a member of a home-base team and one expert group. Ask the participants to move into the expert groups.

3. Hand out the three readings, one topic to each expert group. Ask the expert groups to read and discuss the article about their topic. Each expert group then agrees on key points and decides how to convey these points to their home-base teams.

4. Participants move back to their home-base teams, and each person teaches the home-base team about the topic discussed in his or her expert group. In this way all participants learn about and discuss ideas from all three readings.

5. Circulate among teams to check for understanding.

Activity 2 Planning and Presenting for Adult Learners

After the jigsaw activity, facilitate a whole-group discussion in which statistics educators apply the jigsaw readings to workshop planning. Statistics educators need their workshop plans in front of them.

On newsprint or the board, make three column headings and record key ideas, as illustrated in the chart on the next page.

Characteristics of Teachers as Learners	*Workshop Plans*	*Presentation Skills*
Example: *learn by doing*	Example: *get teachers into an investigation within first 20 minutes*	Examples: *don't intervene;* *let teachers pose own questions*

Provide statistics educators with Handout 3.4, on which they can write notes during the discussion.

What is a characteristic of adult learners that we want to pay attention to during workshops?

Example: Adults learn by doing.

How can we design a workshop to meet this need of adult learners?

Example: We can get teachers involved quickly in an investigation.

What presentation skills will help us meet this need?

Examples: We can let teachers pose their own questions and work toward their own answers. We shouldn't intervene too early.

A point to make here is that Teach-Stat workshops provide teachers with models of active learning. Statistics educators will model teaching methods that promote active learning, so that participants in workshops experience firsthand the process of statistical investigation.

Continue to identify other characteristics of adult learners, and discuss related workshop plans and presentation skills.

Variation: Team Planning

If statistics educators do not have their own plans to evaluate, an option is to use the sample workshop schedules from *Module 2*, Handouts 2.6 and 2.7.

Activity 3 Team Planning Time

Provide 20 to 30 minutes for workshop-planning partners or teams to meet. During this time they apply principles of adult learning, teacher change, and team building to evaluate their own workshop plans.

Look carefully at the planning you have done so far. Evaluate your plans to see whether or not your Teach-Stat presentations will meet the needs of adult learners and teachers' concerns about change. Will your presentation promote team building? Revise your plans as needed.

Circulate among teams as necessary.

Activity 4 Reflection on One's Own Presentation Style

As we prepare for a workshop, let's think about how we can adapt our presentation styles to what we know about the needs of adult learners and teacher change.

In your journals (or on Handout 3.5) write about these questions:

What presentation skills do you have that will be especially appropriate in a workshop with teachers? What presentation skills do you still need to practice? What do you think would help you to develop your presentation skills further?

Allow 7 to 10 minutes of quiet writing time.

Share and discuss the written responses. As they think about ways to develop their presentation skills, some statistics educators may indicate a desire to videotape or audiotape themselves in a rehearsal. Others may ask for a colleague to listen to them practice and to give suggestions. Tell statistics educators that in *Module 7* they may do at least one activity to develop presentation skills.

If you have resource books about presentation skills, you may wish to loan them to statistics educators to read at home.

Conclude the activity on a positive note.

What presentation skills do you already have?

Remind educators that they are already accomplished presenters in many ways. For example, they are already quite practiced in giving clear directions and in keeping track of time. Most of all, they are enthusiastic about Teach-Stat and have had classroom experience with Teach-Stat.

While presenting to one's peers can be intimidating, one significant advantage you will have in this workshop is that you will be teachers helping teachers. You will not likely be perceived as outsiders. You will also have evidence of Teach-Stat investigations from your own classrooms.

Reference

Schifter, D., and C. T. Fosnot. *Reconstructing Mathematics Education.* New York: Teachers College Press, 1993.

Notes

To improve her presentations, one statistics educator decided to time her introductions to make sure she wouldn't "talk to" teachers for long periods of time.

One staff developer defuses statistics educators' anxiety by showing them graphs produced in elementary classrooms or by telling a story about a successful investigation in a school. This reminds statistics educators about their initial enthusiasm for Teach-Stat and puts the focus back on children and classrooms. If the examples are from statistics educators' classrooms, all the better. By recalling their own classroom successes, statistics educators often regain confidence and sense of purpose.

When Teachers Learn

When teachers begin to think in new ways about mathematics, they may first focus on methods and materials. Even more important but less visible changes occur, too. Along with new methods come new attitudes and new knowledge about statistics, new beliefs about how children learn mathematics, and new definitions of the teacher's roles.

Teacher change is about more than materials and activities. Change is about attitudes, beliefs, roles, and understanding of content.

Most Teach-Stat teachers report that their personal understanding of statistics becomes enriched and meaningful for the first time. Teachers who may have excellent computational skills are now beginning to understand what Eisenhart et al. (1993) call conceptual knowledge about "the interconnections of ideas that explain and give meaning to mathematical procedures" (p. 9).

I was scared. I couldn't imagine getting statistics down to an elementary level. Plus, I always had taught math out of a textbook. Now I see that statistics is all around us. The kids learn so much by doing the investigations that I haven't got the books out yet this year.

As teachers learn how to use Teach-Stat investigations in their classrooms, they may change their view of the teacher's role. They may begin to see themselves as facilitators and coaches in a *process* of statistical investigation, rather than as transmitters of mathematical knowledge. Such change means that teachers reflect on basic questions, such as "What does it mean to teach or learn mathematics?"

Because teacher change occurs in attitudes, beliefs, roles, and conceptual understanding of mathematics, it may take months and even several years (Schifter and Fosnot 1993; Wood and Thompson 1993). It is helpful to think of a Teach-Stat workshop as one part of a larger process of change for participants rather than to expect overnight transformations.

How can we facilitate change like that just described? How can we nurture teachers' mathematical knowledge as well as influence how they teach? What do researchers say about the most effective strategies for staff development?

The following summary of research about teacher change can help us make intelligent decisions as we plan workshops.

- Early in the workshop the presenters can help teachers to see connections between Teach-Stat goals and the goals of the teachers, their principals, and schools. For example, with teachers trying to implement the NCTM Standards, presenters can introduce Teach-Stat as a way to meet these standards. Teach-Stat also may be related to goals of interdisciplinary, critical-thinking, and problem-solving curricula. Some teachers commit themselves to Teach-Stat when they recognize content similar to that needed by their students to succeed on statewide tests. Whatever the reason, teachers put more effort

Teachers need to believe that what they are learning is useful in the real world of their classrooms. Teachers ask, *How is Teach-Stat related to the curriculum I am accountable for teaching?*

The teachers who taught the institute had worked the investigations in their own classrooms. They brought graphs their kids had made, and one brought a child's journal to share. We realized it was something we could go back home and use, too.

Adults learn by doing,

by exploring their own theories, even when "wrong,"

by investigating adult issues,

by connecting new learning with prior knowledge,

by interacting with a facilitator,

through freedom from external evaluation,

through some ownership of the workshop,

and by sharing and cooperative learning.

I've picked up the statistics [knowledge] that is relevant to me. If I had to take a statistics course, it just wouldn't seem relevant.

Daily dialogue journals and statistics portfolios have been used successfully in some Teach-Stat workshops to document changes in teachers' attitudes and conceptions.

into learning and changing when they believe in the usefulness of the new ideas (Wood and Thompson 1993).

- Adults learn by doing. Active participation in statistics investigations, followed by reflection, talk, and sharing, is more effective than listening to lectures or watching demonstrations.
- One way in which teachers learn is to be allowed to explore their own theories, even when these turn out to be "wrong."
- Adults learn by engaging in problems about adult issues. This suggests that the investigation topics used in workshops should be about the teachers themselves and their interests.
- Adults have a lower tolerance for not understanding than do children. Thus it is important to help the adult learner to relate new information to what he or she already knows. Teachers, for example, may ask how a new method fits in with or differs from their current curriculum and testing program.
- Adult learners are most comfortable when the presenter acts as facilitator or coach rather than as expert. Adults need opportunities for individual interaction with workshop presenters.
- Anxiety and stress inhibit learning, especially in adults. Adult learners, for example, can be intimidated by formal grades. Grading systems should be avoided in workshops where teacher change is the goal. If you need to account for the performance of the workshop participants, use a checklist of criteria.
- Programs that provide for sharing and collaboration among teachers are more successful than programs without interaction among the teachers (Wood and Thompson 1993). Reflection and metacognition can result from cooperative activity.

References

Eisenhart, M., H. Borko, R. Underhill, C. Brown, D. Jones, and P. Agard. "Conceptual Knowledge Falls Through the Cracks: Complexities of Learning to Teach Mathematics for Understanding." *Journal for Research in Mathematics Education,* 24 (1), 8–40 (1993).

Professional Standards for Teaching Mathematics. Reston, Virginia: National Council of Teachers of Mathematics, 1991.

Schifter, D., and C. T. Fosnot. *Reconstructing Mathematics Education.* New York: Teachers College Press, 1993.

Wood, R. H., and S. R. Thompson. "Assumptions About Staff Development Based on Research and Best Practice." *Journal of Staff Development,* 14 (4), 52–57 (1993).

Expressions of Change

Change is a highly individualized process in adults. As a workshop leader, you will notice that participants ask a variety of questions and progress at different rates.

One model of teacher change is the CBAM or Concerns-Based Adoption Model, in which seven stages of teachers' concerns about innovation are described (Hall 1979; Hord et al. 1987). If you listen carefully to questions that teachers ask in Teach-Stat workshops, you may be able to locate their questions on this developmental model. If you are preparing for a workshop, this model may help you to anticipate concerns of your participants.

One experienced teacher has said, *I see professional development as a very personal process where I determine areas where I want to increase my understanding and abilities* (Raymond, Butt, and Townsend 1992, p. 152).

Stages of Concern: Typical Expressions of Concern About the Innovation

	Stages of Concern		Expressions of Concern
IMPACT	6	Refocusing	I have some ideas about something that would work even better.
	5	Collaboration	I am concerned about relating what I am doing with other instructors.
	4	Consequence	How is my use affecting kids?
TASK	3	Management	I seem to spend all my time getting material ready.
SELF	2	Personal	How will using it affect me?
	1	Informational	I would like to know more about it.
	0	Awareness	I am not concerned about it (the innovation).

(Hord, S. M., W. L. Rutherford, L. Huling-Austin, and G. E. Hall. *Taking Charge of Change.* Alexandria, Virginia: Association for Supervision and Curriculum Development, 1987, p. 31. Reprinted by permission.)

In the early stages of learning about Teach-Stat, teachers will most likely have personal concerns about how Teach-Stat is similar to or different from what they are now doing. Questions raised in this stage are "How will I have to change to do this?" and "How will this affect me?" At this stage teachers also want more information about Teach-Stat—where it originated, who endorses it, what preparation teachers need to use it, and how they might be evaluated when they use it.

Concerns about self:

How will using Teach-Stat affect me?

How will I find time to get all those materials ready?

How will I have to change in order to do Teach-Stat?

Sometimes the personal concerns at this early stage appear as resistance to change; teachers may "characterize the innovation as nothing new, but as something they have always done or used to do. . . . They may convince themselves they really do not have to change" (Hord et al. 1987, p. 31). It is helpful for facilitators to interpret this reaction as a symptom of anxiety and not as a sign of hostility!

Concerns about managing the change:

How can I arrange my classroom schedule to create more time for statistical investigations?

When should I use a stem plot?

According to the CBAM model, as personal concerns are satisfied, teachers' questions become *how-to* questions. Concerns at this how-to stage are about finding and organizing materials and planning investigations. In Teach-Stat workshops, teachers also want to clarify their own content knowledge about statistics and graphs. The how-to questions, then, are about how to *do* the statistics as well as about how to *teach* the statistics.

Concerns about impact:

What is Teach-Stat doing for my students?

I'd like to collaborate with a teacher in another grade level.

I have an idea for making the Raisins *investigation work better for my kids.*

Later stages in the CBAM model are most likely to be evident in teachers who have already used some Teach-Stat approaches in their classrooms. These teachers talk about their students' thinking and the effects of certain strategies on students. Teachers at this stage ask about ways to integrate Teach-Stat with other content areas. They may suggest variations and improvements for activities presented in the workshop, all of which shows that they are making these activities their own and integrating them into their own planning. Teachers at this stage especially benefit from sharing ideas with each other. Teach-Stat follow-up sessions often include extended time for teachers to share.

The CBAM model, whether or not we agree with it, reminds us to listen carefully to workshop participants. Participants' questions and concerns can guide us as we help them make the process of statistical investigation their own.

References

Friel, S. N., and J. H. Gann. "Implementing the *Professional Standards for Teaching Mathematics*: Making Change in Schools." *Arithmetic Teacher* 40: 286–289 (1993).

Hall, G. E. "The Concerns-Based Approach for Facilitating Change." *Educational Horizons* 57: 202–8 (1979).

Hord, S. M., W. L. Rutherford, L. Huling-Austin, and G. E. Hall. *Taking Charge of Change*. Alexandria, Virginia: Association for Supervision and Curriculum Development, 1987.

Raymond, D., R. Butt, and D. Townsend. "Contexts for Teacher Development: Insights from Teachers' Stories." In A. Hargreaves and M. G. Fullan (eds.), *Understanding Teacher Development*. New York: Teachers College Press, 1992.

Team Building

Team spirit among workshop participants provides a context in which teachers can take the risks necessary for learning a new topic such as statistics and for changing how they teach mathematics. Within a workshop team each teacher must feel secure enough to say, "I don't understand that yet," or "I have an idea I'd like to get your reaction to," or "This is not like anything I've ever done before," or "I'm scared." A goal of Teach-Stat is to create a network of teachers who will continue to support each other after they return to their classrooms.

When I told my group how nervous I was about statistics, they talked to me. We built each other up and helped each other. They said that they would be there for me and that I didn't have to worry.

Teamwork is not fully utilized by many classroom teachers. This is not surprising when we consider that successful teachers need to work well on their own during most of the school day (Maeroff 1993). Maeroff contends that "not only do teachers seldom collaborate, but they are not expected to be either leaders or followers of other teachers" (p. 514). Even in schoolwide workshops where participants know each other, they may not yet be comfortable enough to take risks within the group.

Team building goes beyond the familiar warm-up or get-acquainted activities used at many inservice workshops. Ideally, team building occurs in a site away from the usual workplace and includes shared meals and recreation as well as academic challenges. Obviously, the time allotted to team building depends on available resources. Even small events, such as eating lunch together during a one-day workshop, contribute to the bonding of group members. Some workshop facilitators have given participants T-shirts and buttons to promote a sense of group commitment. When time permits, a group session to explore results of a learning styles inventory can help participants to appreciate strengths and styles within the team.

I was scared. I thought I would not understand what was going on. I have to get to know people and to feel comfortable before I can even think. After the first day I was OK.

Two aspects of Teach-Stat workshops serve to build teams. First is group problem solving, which is so essential to the success of statistics investigations. Second is the use of investigation topics that promote sharing of personal information. For example, *About Us*, an introductory investigation, helps to move workshop participants quickly into statistics even as they learn more about each other.

Reference

Maeroff, G. I. "Building Teams to Rebuild Schools." *Phi Delta Kappan,* 74 (7), 512–19 (1993).

Planning for Adult Learners

This chart can help you to apply ideas from the readings and discussion about needs of adult learners and teachers' concerns about change. As you use the chart, focus on the question, "How can I design my workshop to address the needs of adult learners and teachers' concerns about change?"

Needs of Adult Learners and Teachers' Concerns About Change	Workshop Plans	Presentation Skills
Example: learn by doing	Within first 20 minutes begin an investigation.	I'll let teachers pose their own questions. I'll intervene only if a group is silent and stuck for a while. Then I'll ask a question. I won't take over.

My Presentation Skills

1. What presentation skills do I have that will be especially appropriate in a workshop with teachers?

2. What presentation skills do I still need to develop?

3. How can I improve my presentation skills?

MODULE 4

Issues in Teaching Statistics

Overview

In *Activity 1, Self-Assessment of Content Knowledge,* statistics educators assess their understanding of statistics and probability. In *Activity 2, Review of Local Curricula,* they analyze the statistics content in local mathematics curriculum guides. By the end of *Activity 3, Clarifying Content Questions,* statistics educators generate a list of concepts and procedures they need to study further.

Statistics topics listed by statistics educators as needing further study can be addressed in *Module 6, Teaching Vignettes,* along with other content knowledge that you, as the staff developer, identify as important for the workshop. If you plan to ask a statistician to help you in *Module 6,* the list generated in *Module 4* can guide that consultant.

Goals

Statistics educators will

- assess and begin to extend their content knowledge in statistics and probability
- identify statistics topics in local curriculum guides
- analyze the content knowledge needed to facilitate Teach-Stat investigations

Materials

- *Self-Assessment* (Handout 4.1)
- *Teach-Stat for Teachers: Professional Development Manual*
- *Investigation Content Analysis Chart* (Handout 4.2, two copies for each statistics educator)
- *Concept Map of the Process of Statistical Investigation* (Handout 4.3)
- *Investigation Matrix* (*Module 2,* Handout 2.3)

Notes

Activity 1: 90 minutes

Activity 2: 30 minutes

Activity 3: 10 minutes

- statistics reference books
- local mathematics curriculum guides
- *Curriculum Review* (Handout 4.4)

Facilitating *Module 4*

Because the purpose of this program is to prepare statistics educators to teach statistics topics for teachers of all grades from 1 to 6, this self-assessment of content knowledge may produce some anxiety. You, the staff developer, will need to provide a supportive environment while challenging statistics educators to extend their understanding of statistics concepts.

Activities 1, 2, and *3* are guided by handouts. Statistics educators may work in pairs. By clarifying questions about content with each other, statistics educators practice talking about statistics with adults and identify their own level of understanding of statistics concepts.

This module may be introduced by establishing a rationale for the statistics educators to learn even more statistics. One way to begin is by telling a brief story about a personal experience in which you were asked challenging questions during a staff development workshop. The point of the story should be that workshop presenters must continually review and learn more about their topics.

Activity 1 Self-Assessment of Content Knowledge

Why do we need to learn more statistics?

Although it is OK not to know everything about statistics, as a Teach-Stat workshop facilitator, you want and need to know more about statistics than the workshop participants know. Why is it important for workshop facilitators to have a deeper and richer knowledge of content than the participants attending the workshops?

Possible responses are below.

- Our audience will be trying to relate statistics to other content areas and units they teach. We will need to be able to help them see connections and applications.
- We will have adult learners who may ask more advanced questions than children.

- We want to teach content that will help our elementary students when they get into middle school and high school. What we teach has to fit with what students will learn later.
- Knowing more than the audience knows about the content gives us self-confidence for facilitating workshops.

Points that may be used to extend this discussion follow.

Does this mean that we need to know high school or college-level statistical concepts?

There is some debate among educators about how far beyond the students any teacher needs to be prepared. What exactly does it mean to know "more" statistics than somebody else?

Note that because statistics educators may be asked to lead workshops for teachers of any grade from 1 to 6, they should at least understand all of the statistics and probability concepts expected of a grade 6 student. They will want to be prepared for teachers' questions that extend into content beyond that of the elementary-school curriculum. Because statistics educators have already completed a regular Teach-Stat workshop, the necessary concepts should be familiar. Even so, statistics educators may feel uncomfortable about their level of content knowledge.

In activities that follow, statistics educators reflect on their own understanding of statistics and probability and make a plan for extending their knowledge. This self-assessment occurs in the context of planning for a Teach-Stat workshop. Even teachers experienced in teaching statistical concepts to their students may be anxious about statistics. As a staff developer, an important role is for you to establish a sense of safety and freedom in which statistics educators ask questions about topics they do not fully understand.

Deciding what you, as a statistics educator, need to know in terms of content before you facilitate a Teach-Stat workshop is a highly individualized task. For example, what [Debbie] knows about probability is different from what [Jose] knows. Therefore, whatever additional understanding [Debbie] needs to develop differs from the additional understanding [Jose] needs.

This next activity is designed to help you with two things. First, it will help you assess what content you need to know to facilitate the activities you have chosen for the Teach-Stat workshop. Second, it will help you make a plan for extending your understanding of statistical concepts.

Materials

The statistics content knowledge addressed is that needed for the upcoming workshop presentations. It will be especially important to have available statistics reference books and local mathematics curriculum guides. Reference books for *Activity 1* might include those listed below.

Moore, D. S. *Statistics.* 3d ed. New York: W. H. Freeman and Co., 1991.

Landwehr, J., and A. Watkins. *Exploring Data.* Rev. ed. Palo Alto, California: Dale Seymour Publications, 1995.

Newman, C., T. Obremski, and R. Scheaffer. *Exploring Probability.* Quantitative Literacy Series. Palo Alto, California: Dale Seymour Publications, 1987.

Graham, A. *Statistical Investigations in the Secondary School.* New York: Cambridge University Press, 1987.

Hand out *Self-Assessment* (Handout 4.1), *Investigation Content Analysis Chart* (Handout 4.2) (two copies for each statistics educator), and *Concept Map of the Process of Statistical Investigation* (Handout 4.3).

Explain the directions for the self-assessment, which are described in Handout 4.1. Note that in step 7 on this handout, statistics educators are encouraged to develop their own concept maps to represent the process of statistical investigation. One person's concept map of the investigation process is shown on Handout 4.3 and may be used as a resource but should not be viewed as the only possible map.

Activity 2 Review of Local Curricula

The purpose of *Activity 2* is to correlate topics from Teach-Stat investigations with local mathematics curricula. Directions for how statistics educators will review local curricula for statistics concepts are given in *Curriculum Review* (Handout 4.4), which is handed out now.

Activity 3 Clarifying Content Questions

Even after *Activities 1* and *2*, statistics educators may still have questions about content. After they have worked through *Activities 1* and *2*, develop a list of the terms, concepts, and procedures still needing clarification or instruction. Use one or more of the following options to address their content questions:

- Ask for volunteers to serve as content experts on certain topics at your next meeting.
- Prepare minilectures on these topics.
- Select vignettes from *Module 6* that address relevant statistics concepts.
- Enlist the aid of outside consultants, such as a statistician, to meet with your group to answer questions.

Self-Assessment

This activity may be done independently or with a partner.

1. Choose one Teach-Stat investigation you will be presenting, preferably the investigation about which you feel least comfortable. Put the name of the investigation at the top of the *Investigation Content Analysis Chart* (Handout 4.2).

2. Use the *Investigation Matrix* (Handout 2.3) to identify key topics presented in your investigation. Write key terms, concepts, and procedures in the left column of the chart.

3. Using the *Teach-Stat for Teachers: Professional Development Manual,* review the investigation itself to add other key vocabulary, concepts, and procedures to the chart.

4. Ask a colleague to review your chart to see if anything has been overlooked. Revise your list as necessary.

5. Look up important concepts and procedures in a statistics reference book, and read about your topic.

6. Decide how key terms, concepts, and procedures of this investigation are related to each other and to the whole process of statistical investigation. To do this you may find it helpful to use a graphic, an outline, or a concept map. One person's concept map of the process of statistical investigation may be found in Handout 4.3, but this is only one possible map of the process.

7. For each investigation that you will facilitate, repeat steps 1 to 6.

8. Make a list of terms, concepts, or procedures that are still unclear to you.

Investigation Content Analysis Chart

Investigation Title ______________________________

Key Terms/Concepts/ Procedures	*Reference*	*Notes*

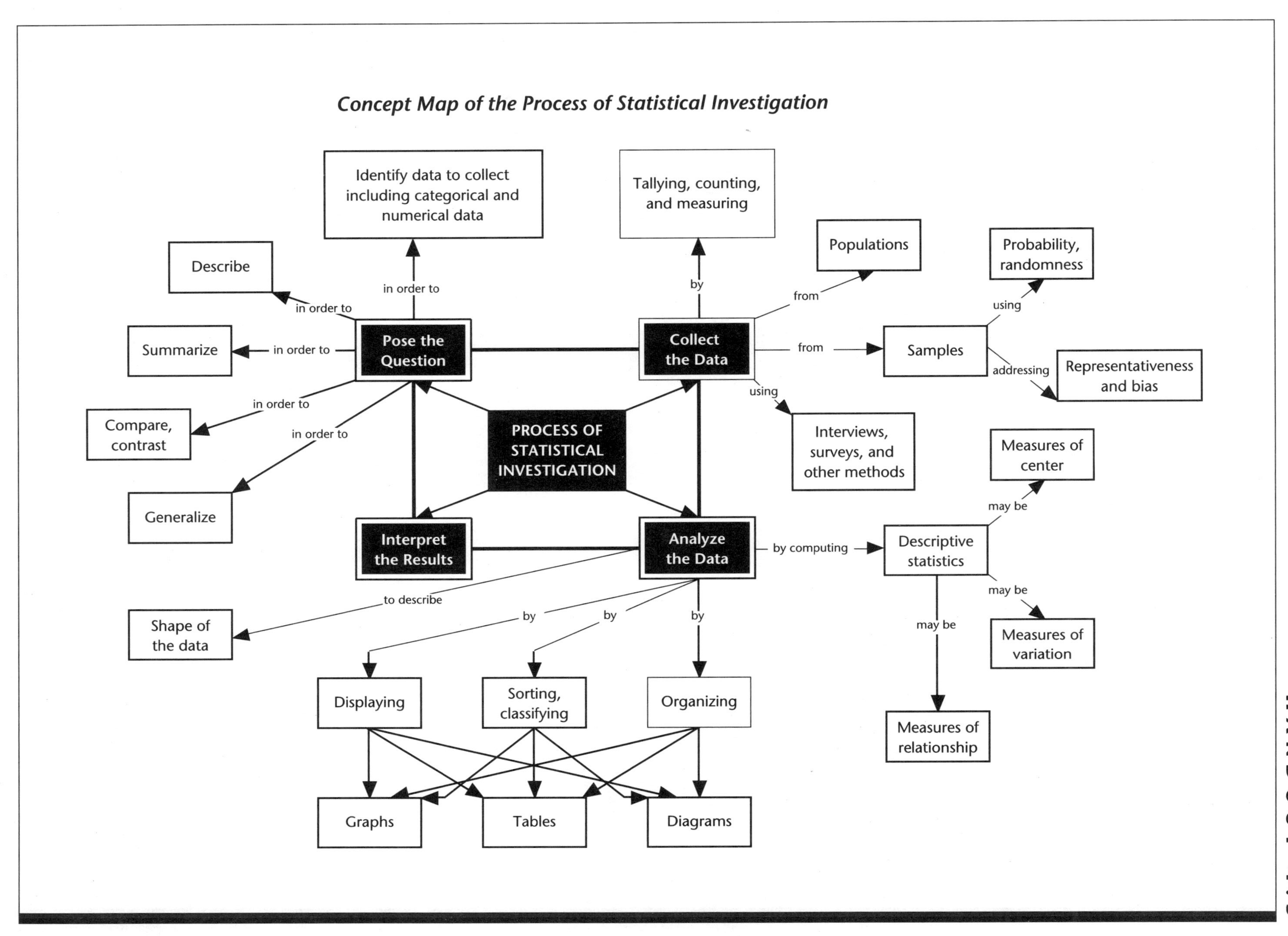
Concept Map of the Process of Statistical Investigation
Identify data to collect including categorical and numerical data
Tallying, counting, and measuring
Describe
in order to
in order to
by
Populations
from
Probability, randomness
using
Pose the Question
Summarize
in order to
Collect the Data
from
Samples
addressing
Representativeness and bias
using
in order to
Compare, contrast
in order to
PROCESS OF STATISTICAL INVESTIGATION
Interviews, surveys, and other methods
Measures of center
may be
Generalize
Interpret the Results
Analyze the Data
by computing
Descriptive statistics
may be
to describe
Shape of the data
by
by
by
may be
Measures of variation
Displaying
Sorting, classifying
Organizing
Measures of relationship
Graphs
Tables
Diagrams

Curriculum Review

1. Obtain a mathematics curriculum guide for the school system in which you will be conducting your workshop.

2. Identify the statistics and probability objectives for each grade level represented by the teachers who will attend your workshop.

3. Identify the statistics and probability objectives covered at the grades one or two levels beyond those grades your participants teach.

4. Compare the curriculum objectives with the concepts and procedures addressed in the Teach-Stat investigations you will be facilitating. For example, where do concepts and procedures of your investigations fit into curriculum used by the teachers who will attend the workshop?

5. Note any topics from the curriculum that you want to study further before the workshop.

MODULE 5

Planning the Details

Overview

In *Activity 1, Refining the Workshop Agenda,* statistics educators look again at the statistics content and format of the workshop agenda they proposed in *Module 2* and make revisions.

Activity 2, Assuming Responsibilities, clarifies job descriptions for the presenting, administrative, and informal roles that occur during workshops. Statistics educators assume responsibilities for all roles involved.

During *Activity 3, Planning for Participant Feedback and Evaluation,* formative and summative evaluation strategies are incorporated into the workshop schedule. Daily comment or feedback cards are discussed.

In *Activity 4, Ordering Supplies and Materials,* statistics educators prepare lists of materials needed in investigations for which they are responsible.

Goals

Statistics educators will

- finalize the agenda and schedule for the workshop being planned
- specify materials they will need during the workshop, including materials that need to be photocopied for participants' packets
- plan for the workshop participants' ongoing feedback and final evaluations
- assume a variety of roles for the workshop

Materials

- copies of the agenda or workshop schedule planned by statistics educators in *Module 2*
- *Concept Map of the Process of Statistical Investigation* (Handout 4.3)

Notes

Activity 1: 30 minutes

Activity 2: 30 minutes

Activity 3: 30 minutes

Activity 4: 45 minutes

Statistics educators often prefer to divide responsibilities for the workshop *before* they review statistics content. This sequence, which is carried out in *Modules 5* and *6*, enables statistics educators to begin to plan their presentations as they review statistical concepts.

Even skeptical staff developers, who may have worried at first about giving so much responsibility for presenting to new statistics educators, have been surprised with the effectiveness of statistics educators during Teach-Stat workshops. The power of teachers teaching each other can overcome inexperience in presenting.

Having a team of workshop presenters is an advantage in terms of the support offered among statistics educators.

Roles of the staff developer

coaching
troubleshooting
recordkeeping
ordering materials

- *Teach-Stat for Teachers: Professional Development Manual*
- *Investigation Matrix* (Handout 2.3)
- *Checking It Twice* (Handout 5.1)
- *Who Will Do What?* (Handout 5.2)
- *Formative Evaluation* (Handout 5.3)
- *Sample Summative Evaluation Form* (Handout 5.4)
- *Workshop Supplies and Materials* (Handout 5.5)
- *Activity Supplies* (Handout 5.6)

Facilitating *Module 5*

This is a critical time in statistics educators' preparation. Making lists, signing up for responsibilities, and talking about evaluation sometimes lead to anxiety. Your first role as staff developer is to reassure statistics educators. Anxiety is a normal physical reaction that helps us make the energy we need to get ready for new events. Emphasize the support within the group of statistics educators: each investigation leader will have a partner, and all statistics educators will be on duty during the entire workshop to assist each other.

As staff developer, you may need to reassure yourself that statistics educators will come through with an effective workshop. Try hard not to take over the presenting roles of the workshop, even if you are asked to by statistics educators. It may help to think of your role as that of a coach or supervisor of student teachers; you will help your team get ready, but you cannot play the game or teach the classes for them!

What you *can* do for the presenters is to check the workshop agenda planned in *Module 2* for potential problems.

- Is the sequence of events logical?
- Does any half-day period seem too full or not full enough?
- Are homework assignments realistic for time frames and audience?
- Are orders for materials complete for every event on the schedule?
- Is each statistics educator signed up for a fair share of the tasks, including delivering presentations?
- Are there any presenters who may need extra help from you?

As you review the agenda, discourage trading of presentation assignments. Sometimes a statistics educator feels most comfortable being the administrative aide every day and avoids the presenting roles. Make it clear that you expect every statistics educator to present or facilitate an investigation.

Another role of the staff developer is that of recordkeeper and, depending on the situation, that of ordering materials. You will probably find it helpful to hand out a list that updates who is responsible for each event and the status of materials ordered.

In sum, as you approach the beginning of the workshop, expect some anxiety, but try to remain calm, optimistic, and reassuring.

Materials

Two useful resources about workshop leadership:

Bowers, P. S. "The Nuts and Bolts of Planning Workshops for Teachers." *Science Educator,* 3 (1): 27–31 (1994).

Sharp, P. A. *Sharing Your Good Ideas: A Workshop Facilitator's Handbook.* Portsmouth, New Hampshire: Heinemann, 1993.

Activity 1 Refining the Workshop Agenda

This set of exercises will help statistics educators to fine-tune the workshop agenda they originally planned in *Module 2.* Common problems at this point are

- too many investigations
- no planned time for workshop participants to ask questions, reflect, and take notes
- poor sequence of investigations
- individual ownership of topics, leading to reluctance to modify the original agenda
- statistics educators' tendency to focus only on the investigations they will lead rather than on the agenda as a whole

Notes

Effective workshops usually use one or two investigations for each half day of workshop time. Although groups vary, most complete an investigation in 45–60 minutes.

Exercises are described here for helping statistics educators to revise the workshop agenda. Choose any of these, depending on the needs of your statistics educators.

Maybe We'd Better Look at the Map

Hand out the workshop agenda that the statistics educators planned during *Module 2.* Statistics educators will also need the *Concept Map of the Process of Statistical Investigation* (Handout 4.3) and several copies of the *Teach-Stat for Teachers: Professional Development Manual.*

Use *Maybe We'd Better Look at the Map* if you believe the statistics educators need to *add* statistics content to their workshop schedule.

We want to make sure we have planned an appropriate survey of statistical topics and concepts for our workshop. Let's use the concept map (Handout 4.3) to mark those statistical concepts we have included in our proposed workshop agenda.

This activity not only reminds the group about the agenda but also helps statistics educators to view the planned workshop as a whole made of related parts.

Is there anything we are leaving out?

Tracing the map can lead to discussion about what statistical concepts the proposed agenda has omitted.

Do we need to incorporate these omitted concepts in other parts of our agenda? If so, how will we do that?

Statistics educators may decide, for example, to teach about categorical data by adding that topic to the *Types of Data* activity rather than by using the *Cats* investigation.

Checking It Twice

Hand out *Checking It Twice* (Handout 5.1).

Let's make sure we are not inadvertently omitting important events by checking our agenda against these criteria. For example, when in our workshop will we gather feedback from the teachers, and when will we respond to that feedback?

Revise the workshop agenda as necessary.

In a separate place, statistics educators record administrative tasks that remain to be discussed or assigned. This list might contain items such as attendance records for continuing education credit, preparation of participants' packets, and photocopying.

Notes

If you think statistics educators still need to incorporate daily feedback, snacks, and administrative tasks to their agenda, use *Checking It Twice* (Handout 5.1).

	M	T	W	T	F
A.M.					
P.M.					

A large chart of the agenda helps to track who is doing what. Having this chart will alert you, as the staff developer, if any statistics educator is assuming too many or too few duties.

Activity 2 Assuming Responsibilities

Hand out *Who Will Do What?* (Handout 5.2).

First, write in the days and times of the investigations you will assist with and facilitate. Make sure that what you write matches the master copy of our workshop agenda.

In a similar way help statistics educators to sign up for remaining roles listed on the handout *Who Will Do What?* Briefly explain each role as described below.

Each morning one of you will welcome the participants and introduce the day's agenda. This may also include presenting our responses to comment cards from the previous afternoon.

Each afternoon one of you will conclude the day by prompting comment cards or journal writing, making announcements, and previewing the next day's events.

The final presenting role is that of distributing and introducing books, software, or other resource materials to the teachers. This involves getting materials to the room, passing them out, giving an overview of what the materials are and how they are useful, and leading teachers through an overview of the materials.

In addition to these presenting roles, there are important assisting roles each day. For example, each of you will sign up to arrange for meals and snacks. Sign up for meal and snack duty on days when you are not the leader or understudy for an investigation.

Administrative aides handle unexpected errands and clerical needs.

After these more formal roles are assigned, discuss functions of statistics educators who are not in a formal role at a given time.

What will the rest of us be doing while other statistics educators are in these formal roles?

Remind statistics educators not to gather in the back of the workshop room to talk during the workshop itself.

For your records, make sure that statistics educators' names are written next to events on the schedule for which they are responsible.

Finally, ask for volunteers to take on any remaining administrative tasks, such as typing materials for participants' packets. During *Module 8* time is allotted for putting packets together, photocopying, and setting up equipment.

Activity 3 Planning for Participants' Feedback and Evaluation

Point out the different functions of final evaluations and daily feedback and the need for both.

The final or summative evaluation, conducted at the end of a workshop, helps us when we plan future workshops. The comment or feedback cards help us to address needs of current participants. Comment or feedback cards also help teachers to feel a sense of directing their own learning.

As a group, identify times for participants to write comment cards and times to respond to the cards. One plan is to gather comment

Notes

Administrative roles during workshops may include preparing participants' packets, keeping attendance records, and gathering journals.

Plan who will act as correspondent to workshop participants. Correspondence may include acceptance letters and welcoming letters with information about the workshop site, times, requirements, and which supplies to bring.

Informal roles include:

sitting among the participants

observing participants in order to diagnose confusion

facilitating small-group work when requested by an investigation leader

At one workshop statistics educators decided they would each sit with a small group of workshop participants rather than sit together at a table at the back of the room. They discovered that this integration with participants led to more attentive participation by everybody. These statistics educators also found they could be more sensitive and responsive to participants' questions.

cards at the conclusion of each day. Statistics educators read and discuss the cards during their daily meetings after the workshop sessions. The statistics educator in charge of introducing on the following day presents responses to the cards as part of that introduction.

Daily dialogue journals have been used successfully as an opportunity for workshop leaders to individually evaluate participants' statistics conceptions. Daily journal entries typically have two parts. First is often a request that participants respond to a "content" or "methodology" question related to the investigation. For example, "Which measure of center might be the best to use to represent the amount of money teachers spend on buying classroom materials? Why?" A second journal entry might ask participants to reflect upon their reactions to the day's events. Because it is often impractical for one person to collect and review every participant's journal on a daily basis, a plan may be devised to collect and review participants' journals on a rotating basis.

Notes

In one workshop, participants were organized into grade-level teams and each statistics educator was designated as a facilitator for a team. Review of daily dialogue journals was then managed by each statistics educator for his or her grade-level team.

To discuss these options for evaluation, you may use Handout 5.3, *Formative Evaluation.*

While giving participants continuous feedback on their learning can be accomplished by using dialogue journals, a more summative assessment of teacher learning or change is often also desirable. Portfolios have been used for this purpose in Teach-Stat workshops with varying degrees of satisfaction. At some Teach-Stat sites, teachers collected evidence to illustrate the changes in their understanding of statistics concepts and methodology by taking photographs of their daily work. The pictures were then organized sequentially into an album with the teachers' editorial comments provided along with each photo. If group research or investigations are part of your workshop, a report of the group project may also be included in one's portfolio. At another workshop, teachers included a summary "letter" to the workshop leaders at the end of each portfolio as a way of reflecting upon their professional development as a result of the workshop. If portfolios are used, teachers may like to share them—either in a small- or large-group setting—on the last day of the workshop.

Other items teachers include in portfolios are printouts from computer software used during the workshop. If teachers use graphing software, for example, they may include in their portfolios data representations generated by that software.

As a staff developer, you will want to ask the statistics educators for their ideas regarding the use of dialogue journals and portfolios for your workshop. If statistics educators decide to use a final evaluation form, design that form now.

When the workshop is over, what will we want to know from the participants? Make a list of what you will want to know.

Distribute copies of *Sample Summative Evaluation Form* (Handout 5.4).

Will any items on the handout help us gather that information? What other items do you want?

As a group, check the workshop agenda to make sure adequate time has been provided for daily and final evaluation.

Make decisions about the format of the final evaluation, including who will be responsible for handing out and collecting forms.

Activity 4 Ordering Supplies and Materials

Hand out *Workshop Supplies and Materials* (Handout 5.5).
As a group, begin with Day 1 of the workshop, and make a list of materials needed. The lists of materials in Handout 5.5 and in the *Teach-Stat for Teachers: Professional Development Manual* will be helpful. Recording the approximate cost of items will help statistics educators keep expenses within budgetary constraints.

Decide who will place, verify, and follow up on orders. Arrange for storing the supplies. It is helpful to label boxes of materials with their contents and quantities.

Checking It Twice

Use this list to refine your workshop schedule.

_____ Teachers have opportunities to get acquainted and to share informally.

_____ The sequence of investigations makes sense conceptually.

_____ Teachers have times for questioning, commenting, and making suggestions.

_____ Opportunity is provided for obtaining information and giving feedback about teachers' conceptual development.

_____ Facilitators have time for responding to teachers' daily concerns, as in responding to comment cards.

_____ Facilitators have time to meet at the end of each day to make continual adjustments to plans and to share feedback from teachers (suggested time: 1 to 1½ hours).

_____ Enough time is allotted to each investigation, including time to question, write notes, and share (most investigations take 45 to 60 minutes).

_____ Homework assignments are realistic (not longer than 45 minutes).

_____ Minilectures, videotapes, or software demonstrations do not follow lunch.

_____ Sufficient time is allotted on the last day for wrap-up and final evaluation.

_____ Recognition of teachers is planned at the conclusion of the workshop.

Who Will Do What?

Statistics Educator's Name ______________________________

Role	*Investigation/ Activity*	*Day*	*Time*	*Notes*
■ Investigation/Activity Facilitator				
■ Understudy to Investigation Facilitator				
■ Introducing a Day				
■ Concluding a Day				
■ Introducing a Material or Resource				
■ Snack Preparation and Setup				
■ Administrative Aide				
■ Other				

Formative Evaluation

Consider the following questions for journals and comment cards:

What ideas that I learned today do I want to remember?

How can I use those ideas in my classroom?

What questions do I still have about these ideas?

What concerns or suggestions do I want to give to the workshop facilitators?

What did I like best about today?

What did I like least about today?

Sample Comment Card

Significant ideas I learned today:

Question or concern I still have:

Suggestions:

Notes

Be sure to provide time at the end of the day for participants to write in their journals or on comment cards. Comment cards are written anonymously. Collect cards in a basket by the door as participants leave for the day.

Journals may also be collected at the end of the day. Written responses to journal entries can provide individualized help for workshop participants.

Sample Summative Evaluation Form

Directions

Rate your agreement with each statement by circling a number from 1 (strongly disagree) to 5 (strongly agree).

	strongly disagree	disagree	uncertain	agree	strongly agree
	1	2	3	4	5

A. Content of Workshop

1. Statistics concepts used in this workshop are relevant to curriculum I plan to teach.
2. Information about how to teach statistics, as presented in this workshop, will be useful in my classroom.

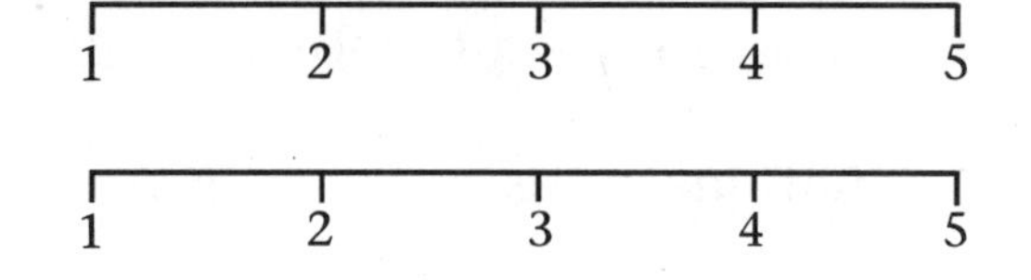

3. Demonstrations and investigations performed in the workshop could be used in my classroom.
4. The workshop was at an appropriate level of challenge for me.

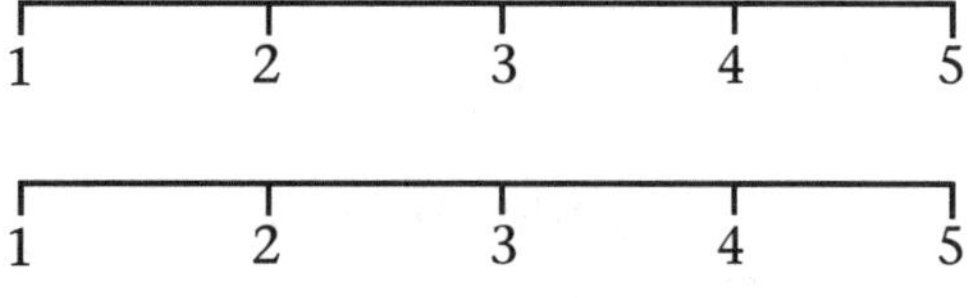

5. Adequate time for sharing and reflection was provided.

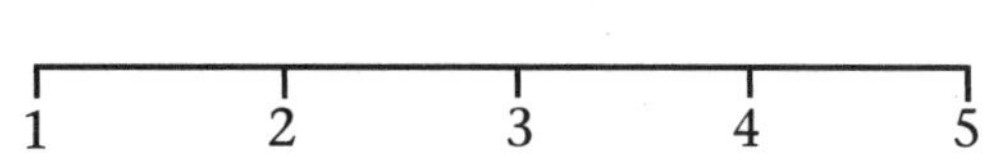

B. Workshop Presentation

1. The pace of the workshop was appropriate for me. 1 2 3 4 5
2. Concepts were clearly presented. 1 2 3 4 5
3. Workshop facilitators responded to participants' questions and needs. 1 2 3 4 5
4. I understand statistical concepts better now than I did before the workshop. 1 2 3 4 5

C. Workshop Site

1. The location of the workshop was convenient.
2. The physical facilities were satisfactory.

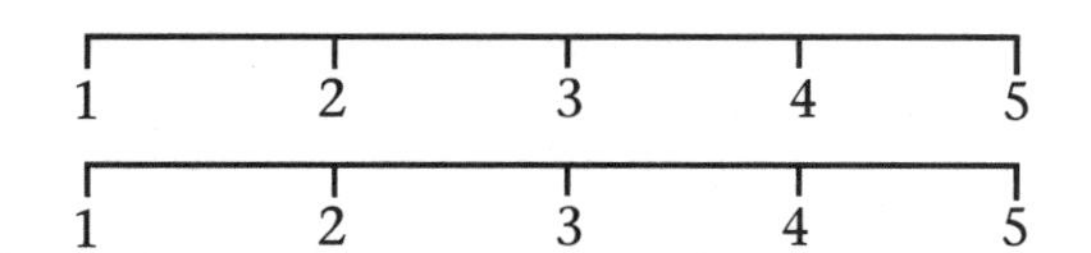

D. General

In general, I feel this workshop was valuable for me.

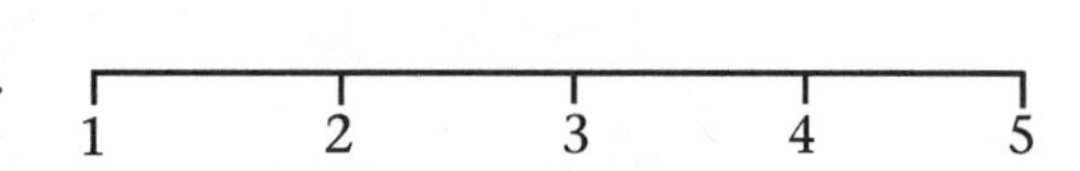

E. Please add your comments in the space below.

Workshop Supplies and Materials

Use this worksheet to keep track of supplies and materials you will need. Quantities given here are either for each group of 4 participants or for each workshop of 24. Some supplies are needed only for certain investigations. Order supplies only after a review of investigations you will use.

Item	*Quantity per 4 people*	*Supplier*	*Cost*	*Person responsible*
supply basket	1			
scissors	2			
stick-on dots in assorted colors	1 pkg. of 1000 (4 pkg. to take)			
stick-on notes, 1½ x 2 inches	500 (1000 to take)			
centimeter-graph paper,	16 sheets			
masking tape	1 roll			
transparent tape	2 rolls			
thumbtacks	1 box			
felt-tip markers, broad-point assorted colors	2 boxes 1 box for presenter			
rulers	2			
yardsticks/meter sticks or tapes	2			
calculators	4			
stopwatches	2			
A/V equipment overhead, screen, VCR, TV chart stand or chalkboard computers, printers extension cords				
Used Numbers videotapes				
transparency film, markers				

Item	*Quantity per 24 people*	*Supplier*	*Cost*	*Person responsible*
chalk or dry-erase markers	1 pkg.			
cameras and film, batteries	2			
name tags	24			
participants' packets list of participants and leaders copy of the schedule copy of requirements list of materials to be received journal or paper for notes copy of the NCTM Standards pencil two copies of all worksheets renewal credit forms	24			
final evaluation forms	24			
index cards for daily comments	24 per day			
blank computer diskettes	1 box			
software				
interlocking cubes	1 pkg. of 1000			
ball of string, yarn	1			
books, such as *Used Numbers*				
large flip charts of graph paper marked in inch squares	200 pages			
flip chart of newsprint	1			
workshop schedule on newsprint	1			
daily attendance sheets	1 set			
classroom clock	1			
welcome/acceptance packets	24			
cups, plates, napkins, hot water pot, extension cord, coolers				
trash bags and trash cans				

Activity Supplies

List supplies needed for specific investigations, lectures, or activities.

Activity	*Item*	*Cost*	*Person responsible*	*Notes*

MODULE 6

Teaching Vignettes

Overview

In *Module 6* statistics educators develop both understanding of statistics and skills in workshop discourse. Key points are the following:

- Conceptions about statistics can develop through classroom discourse about data
- Those developing and sometimes naive conceptions are most effectively viewed by workshop leaders not as errors but as opportunities for learning
- Investigations provide contexts for discourse about data
- Active listening and careful questioning are useful strategies for helping adults as well as children to construct statistics concepts

After a brief introduction in *Activity 1, Introducing Pedagogical Content Knowledge,* statistics educators begin with a vignette about mode in *Activity 2, Introducing the Teaching Vignettes: The Vignette About Mode.* In *Activity 3, Practicing with Vignettes,* eight additional vignettes are used to stimulate classroom discourse. Discussion questions and sample responses follow each vignette.

Goals

Statistics educators will

- listen for and identify children's and adults' naive or developing conceptions about statistics
- teach by asking questions rather than by telling answers and thus develop skill in classroom and workshop discourse
- develop their own understandings of statistics

Notes

Activity 1: 15 minutes

Activity 2: 45 minutes

Activity 3: 45 minutes for each vignette

The 45 minutes estimated for reading and discussing a vignette varies according to statistics educators' prior knowledge.

Topics in the vignettes:

mode
grouped/ungrouped data
median
mean
interpreting scatter plots
unequal-size groups
line graphs
probability
responding to participants' questions

Variation

In some Statistics Educators Institutes, staff developers present and analyze three vignettes as a group. Then statistics educators work on the remaining vignettes independently. Independent work is followed by a whole-group discussion.

Materials

- nine vignettes with discussion questions and sample responses (Handouts 6.1–6.9)

Facilitating *Module 6*

As the staff developer, you may wish to present all nine vignettes, or you may select those vignettes about topics identified in the self-assessments made by statistics educators in *Module 4*. One suggestion is to begin with the vignette about mode. This relatively brief vignette introduces statistics educators to the *process* of analyzing and responding to a vignette, before you present more challenging statistics topics.

Activity 1 Introducing Pedagogical Content Knowledge

Introduce the concept of pedagogical content knowledge, which may be new to some statistics educators.

Before you begin teaching your colleagues about statistics, you will want to work on another kind of content knowledge: pedagogical content knowledge. Pedagogical content knowledge involves identifying the most effective ways to present the content—the best analogies, illustrations, examples, explanations, and demonstrations for making the content understandable to others. Pedagogical content knowledge also includes understanding what makes the learning of specific topics either easy or difficult (Shulman 1986).

Preconceptions people bring with them have important influences on learning. Preconceptions include beliefs, attitudes, and stereotypes, as well as mathematical knowledge. The more practice you have in recognizing the preconceptions held by workshop participants, the better prepared you will be as a workshop facilitator.

As someone who has worked with children to develop their statistical concepts, you already have some knowledge of types of problems students may have in understanding various statistical concepts. For example, if you have taught statistics at grade 5, you may have found that students at this level often have difficulty determining the most appropriate intervals for the axes on a histogram.

If you have had a teaching experience like this, you may have decided that the next time your students use histograms, you will be prepared with examples of different explanations so

that they might be more successful. Teaching experience has increased your teaching repertoire—or your pedagogical content knowledge—with respect to histograms.

Another common area of confusion is about the mean, mode, and median and when to use which one. Many people have simply memorized definitions of these terms and think of them all as measures of central tendency and have little understanding of what they really represent. In Teach-Stat we try to broaden teachers' understanding of mean, mode, and median by using the term typical *when posing questions for investigation. For example, we ask, "What is the typical family size?" By posing questions in this way, we allow students or workshop participants to explore which measure of center best represents their data for the question under investigation.*

Similarly, we have found that certain concepts and skills need special attention when you are facilitating Teach-Stat workshops for teachers. In this module we will prepare ourselves to respond to preconceptions, questions, and confusions that commonly occur in workshops. The vignettes in this module will help you to listen and respond to others' preconceptions. As we gain experience in helping others to extend their statistical concepts, we will become better able to define and to take advantage of teaching opportunities in our workshops and classrooms.

Activities of Module 6 *focus on a series of teaching vignettes—stories that feature teachers interacting with students or that show workshop facilitators interacting with teachers about statistics. As we analyze these interactions, we will also stretch our own understanding of statistical concepts.*

Activity 2 Introducing the Teaching Vignettes: The Vignette About Mode

As a staff developer you may have noticed a common preconception about mode, as illustrated here. The graph below represents the number of raisins in a collection of half-ounce boxes of raisins. The mode is 29 raisins.

Boxes of Raisins

25	26	27	28	29	30
				X	
				X	
				X	
				X	
			X	X	
		X	X	X	
		X	X	X	

Raisins

Notes

Following constructivist beliefs about learning, many staff developers avoid using the word *misconception* when learners are in the process of developing statistics concepts.

Words like *preconception, naive conception, alternative conception,* and *developing conception* are useful for workshop facilitators as they challenge teachers to think more deeply about concepts.

The article "What's Typical?" by Mokros and Russell (1989) contains additional examples of people constructing a concept of average. You may wish to use this article as assigned reading or to share several examples from it with statistics educators.

Mokros, J. R., and S. J. Russell. "What's Typical?" *HANDS ON!* 21 (Spring): 8–9 (1989).

When asked what the mode is in this example, a common answer is 7, which is the number of boxes having 29 raisins. What preconception about mode might lead to an answer of 7? How can a workshop leader most effectively respond to such an answer? These are the kinds of questions you can help statistics educators learn to discuss by using the vignettes in *Module 6.*

To begin *Activity 2,* have statistics educators read silently the first page of *Vignette: Mode* (Handout 6.1). Then help them discuss the question at the bottom of that first page.

Hand out pages 2 to 4 of the vignette. Have statistics educators assume the roles of the teacher, Erica, Shaundra, and John and read the dialogue out loud. Discuss the three questions.

After the discussion, hand out page 5 of the vignette, which presents sample responses.

Select three or four other teaching vignettes to review with the group in the manner suggested for *Vignette: Mode.* Remember to hand out the last page of each vignette, *Sample Responses to Discussion Questions,* after statistics educators have generated their own responses.

Notes

Sample responses to the vignettes include questions that teachers could ask in order to facilitate mathematical discourse.

By working through these vignettes, statistics educators can practice *teaching by asking.*

Activity 3 Practicing with Vignettes

Proceed with your discussion of other vignettes, selecting topics that statistics educators have indicated they need to study. The vignettes may also be used for independent study.

Reference

Shulman, L. "Those Who Understand Knowledge Growth in Teaching." *Educational Researcher* 15 (2): 4–14 (1986).

Vignette: Mode

Overview

The *mode* of a set of data is the data value or data category that occurs most often or with the greatest frequency.

Students are sometimes confused when they are asked to identify the mode of categorical data. They may want to report the mode of categorical data as the frequency or count of the observations for the mode category instead of the category itself. In this vignette, the teacher guides students to use information about the graph to help them clarify their thinking about the mode.

Vignette

Students collected and presented data about classmates' birth months.

Classmates' Birth Months

Jan	Feb	Mar	Apr	May	Jun	Jul	Aug	Sep	Oct	Nov	Dec
					X						
					X						
					X				X		
X				X	X				X		
X				X	X				X		
X		X		X	X		X		X		X
X		X	X	X	X		X		X	X	X

Months of the year

Various measures of center were discussed. When asked what the mode of the data was, Erica was one of several students who answered 7.

Teacher: ***Erica, why do you say that 7 is the mode?***

Erica: ***Because 7 is the highest number.***

Teacher: ***Let's look at the mode again. When we ask for the mode in the data, we want to know something about the data we are graphing. What are we investigating with these data?***

Erica: ***We're finding out when we were born. Seven of us were born in June, more than any other month. So 7 is the mode, right?***

What questions might the teacher ask to help Erica develop her conception of mode?

HANDOUT 6.1

Here is one possible continuation of the discourse between Erica and her teacher.

Teacher: *What have we marked along the horizontal axis of the graph?*

Erica: *Months.*

Teacher: *Right. What question are we asking about months?*

Erica: *Which month has the most of us born in it.*

Teacher: *OK. Will the answer to your question be the name of a month or a number?*

Erica: *A month—June will be the answer.*

Teacher: *So when we talk about the mode of our graphed data, we will be talking about a month.*

Erica: *So June is the mode?*

Teacher: *Why do you think so?*

Erica: *Because more people in our class were born in June than in any other month.*

Teacher: *Yes. And how many people were born in June?*

Erica: *Seven.*

Teacher: *When we ask for the mode, we are asking for the month that occurs most frequently in our data. We are not asking for the number of times that the modal month appears. Our investigation question is about which month, and so the mode will be a month.*

Teacher: *Let's look at some other graphs we have made in the past and see how to talk about the modes.*

Boxes

					X					
					X					
					X					
					X					
					X					
					X	X		X		
				X	X	X		X		
				X	X	X		X		
X		X	X	X	X	X	X	X		X
24	25	26	27	28	29	30	31	32	33	34

Number of raisins

Teacher: *When we ask for the mode of these data, remember what we are investigating. What was the question we asked in this investigation?*

Shaundra: *What's the typical number of raisins in a small box.*

Teacher: *Yes. It's very important to keep your question clearly in mind. So, will the mode of this graph be a number of raisins or a number of boxes?*

Shaundra: *A number of raisins, because that's our question.*

Teacher: *Yes. It seems to help when you ask yourself "the mode of what?" In this case, you are looking for the mode of what?*

Erica: *The mode of the number of raisins. And those are the numbers across the horizontal axis.*

John: *Then is the mode 29?*

Teacher: *Why do you think the mode is 29?*

John: *We asked for the typical number of raisins in a box, and more boxes had 29 raisins in them than any other number of raisins.*

Teacher: *Yes. Does it matter how many boxes had 29 in them?*

Erica: *No, just as long as more boxes had 29 than any other number.*

Teacher:	***Number of what?***
Erica:	***Number of raisins.***
Teacher:	***Erica, please use the mode to answer this question: What's the typical number of raisins in a small box of raisins?***
Erica:	***The mode is 29.***
Teacher:	***Twenty-nine what?***
Erica:	***The mode is 29 raisins, and so the typical number of raisins in a small box is 29.***

Discussion Questions: Mode

1. What was Erica's preconception about mode?

2. What impact did Erica's preconception have on her interpretation of the data?

3. What are some possible interventions when this preconception occurs?

Sample Responses to Discussion Questions: Mode

1. Erica may have equated *mode* with *most frequent,* or she may have equated *how many* with *most frequent.*

2. One significant impact of Erica's conception of mode was that she was unable to answer the investigation question about the most common birth month among her classmates.

3. Effective interventions for Erica were (a) awareness of graph labels, (b) review of the investigation question, and (c) asking "the mode of what?" Asking "the mode of what?" led Erica to think about the subject of the graph and the modal subject.

   The teacher could also have rewritten the line plot as follows:

Classmates' Birth Months

					Jun						
					Jun						
					Jun				Oct		
Jan				May	Jun				Oct		
Jan				May	Jun				Oct		
Jan		Mar		May	Jun		Aug		Oct		Dec
Jan		Mar	Apr	May	Jun		Aug		Oct	Nov	Dec
Jan	**Feb**	**Mar**	**Apr**	**May**	**Jun**	**Jul**	**Aug**	**Sep**	**Oct**	**Nov**	**Dec**

Months of the year

Vignette: Grouped and Ungrouped Data

Overview

When we represent data using cubes or squares on grid paper, one way is to use one tower of cubes or one "tower" of grid squares per person. For example, if we investigate the number of pencils we have with us, one person might make a tower of three cubes to represent three pencils, another tower might have five cubes to represent five pencils, and so on. These data are referred to as *ungrouped* or *raw data*. We have *ungrouped data* when each data value (that is, the number of pencils for one person) has its own representation, such as a tower of cubes. The height of the cube tower or bar on grid paper is determined by the value of the individual data point.

Confusion may arise when we put these same ungrouped data into a line plot or bar graph. Each X on the line plot represents one tower. These data are now called *grouped data*. There is no way to look at the Xs themselves and find out what values they represent; for example, an X over the value of 1 on the horizontal axis looks just like an X over the value of 3. Only in reference to values along the horizontal axis do the Xs take on different values. Also, with grouped data, each person's data are impossible to identify: that is, if two people have three pencils, we cannot tell which X on the graph goes with which person.

When students have to work with several of these representations at once, as in the following vignette, they may become confused.

Vignette

Ms. Sanders: ***Count the number of pets that you have and make a cube tower to show that number.***

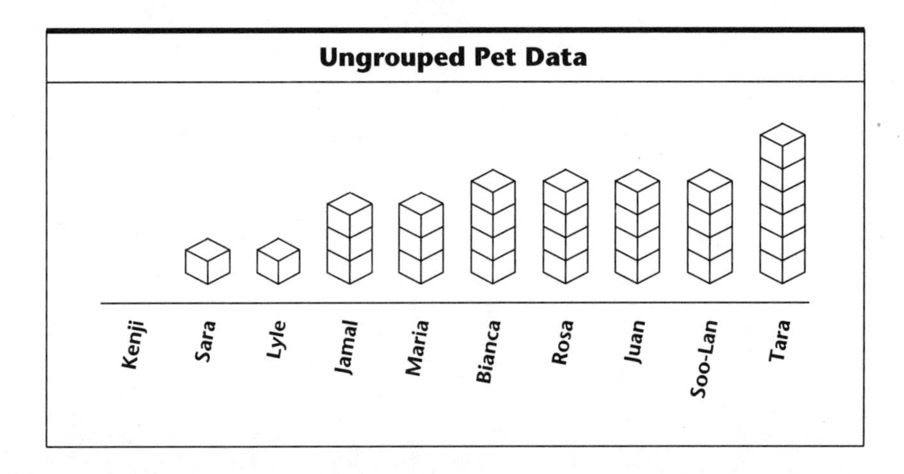

Ms. Sanders: *How could you represent these data with a line plot?*

Pets in Ms. Sanders' Class

				X			
				X			
	X		X	X			
X	X		X	X		X	
0	1	2	3	4	5	6	7

Number of Pets

Ms. Sanders: *What is the mode number of pets in this group of students?*

Lyle: *Well, in the tower graph I think the mode is 6, because that's the tallest tower.*

Ms. Sanders: *What about the mode for the line plot?*

Lyle: *That is 4, because 4 has the most Xs above it.*

Sara: *No, wouldn't it be that 4 is the mode of the tower graph, because there are four towers of 4—more than any other number? And for the line plot, 2 is the mode, because there are two Xs over the number 1 and two Xs over the number 3.*

Ms. Sanders: *Does anyone agree or have a different idea?*

Bianca: *I think the mode is 4 in both the tower graph and the line plot.*

Ms. Sanders: *Why do you think the mode is 4?*

Bianca: *Because more of us have four pets than any other number. That's true in the tower graph and in the Xs in the line plot.*

Discussion Questions: Grouped and Ungrouped Data

1. Only Bianca uses the concept of mode in a statistically appropriate way. How might Lyle and Sara be thinking in order to conclude that the data have one mode in the cube representation and a different mode in the line plot?
2. How do Lyle and Sara's conceptions affect their interpretations of the data?
3. What questions could Ms. Sanders ask that might help Lyle and Sara?

Sample Responses to Discussion Questions: Grouped and Ungrouped Data

1. Lyle may be trying to transfer a strategy that works for grouped data to the ungrouped data. In a line plot or a bar graph of grouped data, the mode is the category having the greatest value or the value that has the tallest "tower of Xs" or tallest bar. This strategy usually does not work for ungrouped data.

   Sara may be doing the opposite, that is, applying a strategy that works with ungrouped data to grouped data. In ungrouped data, such as the cube towers, the mode is the *height* of towers that occurs *most often*. In these data, there were more towers of 4 than towers of any other height.

   These students seem to use a strategy for finding the mode inappropriately. They may not yet understand the difference between grouped and ungrouped data. They might, for example, think about visual aspects of data representations (bars on a graph) without thinking about what those visual aspects represent. They may not consider the type of information represented.

2. If students do not know the limitations of strategies they learn, they are likely to try to apply the strategies indiscriminately. For Lyle and Sara this mismatch between strategies and situations produces incorrect answers. Without an understanding of the errors, students may be unable to correct their thinking about underlying concepts, so that the errors continue to be generated and students develop a conceptual base that inhibits further learning.

3. Ms. Sanders could encourage Lyle and Sara to read data representations aloud.

   - How shall we label this graph?
   - What does each X represent? What does a cube represent?
   - Pick an X. How is this X represented in the cube display?
   - What does a tower of cubes represent?
   - Pick a tower. How is this tower represented in the line plot?

   She could also ask students to define *mode* in the context of the pets investigation. The word *typical* may help students to develop their conception of mode. If the question behind the investigation is clearly understood, this may help students to realize that the value of the mode should be the same value, whether represented by cubes or in a line plot.

   - What questions are we asking about pets?
   - What question can the mode help us to answer?
   - What is the typical number of pets a student in this group has? How many pets does a typical student in this group have? Are these two questions the same or different? Will answers to these questions be the same or different?

Vignette: Median

Overview

This vignette illustrates paths that students may take as they construct their conceptions of median. How many different conceptions and misconceptions of median can you find in the dialogue?

Vignette

When Mrs. English's class investigated the typical number of raisins in a half-ounce box, the following data were recorded:

40, 35, 35, 30, 30, 28, 31, 29, 35, 36, 28, 29, 28, 28, 38, 38, 29, 31, 31, 28, 28, 30, 30, 32, 34, 35, 35, 35, 38, 34

Mrs. English: ***You may recall that we began talking about the median of a set of data yesterday. Sometimes looking for the median helps us decide what is typical. Look at these data. Think for a few minutes about what you think would be the median of these data.***

Shawn: ***It's 15. There are 30 numbers, and half of 30 is 15. So 15 is the middle, or median.***

Paulina: ***No, the median is 38, because there are 30 boxes of raisins and half of 30 is 15, and the fifteenth number is 38.***

Sook Leng: ***I disagree. First you have to put your numbers in order from smallest to largest like this:***

28, 28, 28, 28, 28, 28, 29, 29, 29, 30, 30, 30, 30, 31, 31, 31, 32, 34, 34, 35, 35, 35, 35, 35, 35, 36, 38, 38, 38, 40

Now 31 is the median because there are 30 boxes of raisins and half of 30 is 15 and the fifteenth number is 31.

Erin: ***I almost agree with Sook Leng, but the median has to have the same number of boxes on each side. So, you have to say that the median is between 15 and 16, and that would be 15½.***

Jo Anne: ***I've been thinking about what Sook Leng said. You could think of 31 as the median in another way. I made a line plot. If you look at the tallest towers, they're at 28 and 35. Then you can find the middle of these two towers. That would be 28 + 35 = 63, and 63 divided in half is about 31. So 31 must be the median.***

Jo Anne's Line Plot of the Class Raisin Data

26	27	28	29	30	31	32	33	34	35	36	37	38	39	40
		X							X					
		X							X					
		X		X					X					
		X	X	X	X				X			X		
		X	X	X	X			X	X			X		
		X	X	X	X	X		X	X	X		X		X

Number of raisins

Mrs. English: *Joseph, do you agree that 31 is the median of these data?*

Joseph: *No, I'm thinking the median is 34. See, the smallest number of raisins found was 28, and the largest was 40. If you list these numbers in order, like 28, 29, 30, 31, 32, 33, 34, 35, 36, 37, 38, 39, 40, the middle number is 34, because there are 13 numbers, and you need to find the number with six numbers on each side of it, like six numbers above it and six numbers below it.*

Michael: *I have another idea. The median is 33, because the numbers at the bottom go from 26 to 40. They would balance at 33. There would be seven numbers above 33 and seven numbers below.*

Mrs. English: *I'm not sure about that, Michael. Is 33 the median of the data, or is 33 the median of the scale?*

Michael: *Oh, I see. So 33 would be the median of the scale. But 33 can't be the median of the data, because there aren't any data there! So I guess there isn't a median, or it might be zero since there isn't an X there.*

Cari: *I'm not so sure. I think the median of these data is 32, because when you look at the plot you see that most of the boxes had between 28 and 36 raisins, and the middle of 28 and 36 is 32.*

Brennan: *I agree with Michael. The median is 33, but it's because there are 10 towers of Xs. At 33 we'd have five towers on each side.*

Discussion Question: Median

What conceptions and misconceptions about median are in this vignette?

Sample Responses to Discussion Question: Median

Different conceptions represented in the vignette are listed below.

	Student	Median	Reasoning
1.	Michael	none or 0	no data at point where median is
2.	Shawn	15	median of even number of observations (number of observations divided by 2)
3.	Erin	15½	location of median data point
4.	Sook Leng*	31	attempted to locate median, but did not average two middle data points in an even set of data
5.	Jo Anne*	31	average of both modes
6.	Cari	32	chose middle of a cluster of data
7.	Michael**	33	median of the horizontal scale
8.	Brennan	33	median of the number of columns of data; point where the total number of columns of data would "balance"
9.	Joseph**	34	median of the range of the data
10.	Paulina	38	median of unordered data

*Sook Leng and Jo Anne found the correct median.

**Research has shown that conception 9 was a prevalent alternative among a group of elementary teachers. The next most common conception was 7 (Berenson, Friel, and Bright 1993).

Reference

Berenson, S. B., S. Friel, and G. Bright. "Elementary Teachers' Conceptions of Graphical Representations of Statistical Data." Paper presented at the annual meeting of the Research Council for Diagnostic and Prescriptive Mathematics, Melborne, Florida, February 1993.

Vignette: Mean

Overview

Confusion about the mean is addressed when students consider what is *typical* about their data.

Vignette

Ms. Nuang's class was investigating family size. Students collected data and made a line plot.

Family Size

		X	X						
		X	X						
		X	X						
		X	X						
		X	X						
X		X	X						
X		X	X						
X		X	X						
X		X	X	X	X				X
2	3	4	5	6	7	8	9	10	11

People in family

As students discussed how to describe what was *typical* for their data, mode and median were considered good options for describing typical family size. During the discussion about median and mean, one student, Simone, busily entered numbers into a calculator and finally addressed the class.

Simone: ***I used my calculator and found the mean is 4.52.***

Kristen: ***You can't have 4.52 people in a family!***

Taylor: ***I have a different mean. I added (2 + 4 + 5 + 6 + 7 + 11) and got 35. Then 35 divided by 25 equals 1.40. I know you can't have 1.40 people, but the mean can be a decimal number.***

Kristen: ***How can the mean be 1.40 when there are no families with just one person?***

Discussion Question: Mean

How might you respond?

Sample Responses to Discussion Question: Mean

Suggest that the students use interlocking cubes to show the data. This requires them to work backward from the graph.

Ms. Nuang: ***What does an X on the 2 mean? How would this look if you used the cubes?***

Simone: ***A cube tower with two cubes?***

Kristen: ***And we need four towers of two cubes each to match the four Xs on our graph.***

Taylor: ***So, does that mean we need nine towers of four cubes each to match the nine Xs on our graph?***

Ms. Nuang: ***Yes. Let's make these towers.***

Taylor: ***Okay. I'll make the nine towers of five cubes.***

Simone: ***I've made the four towers of two cubes, so I'll make a tower with six cubes, one with seven cubes, and one with eleven.***

Taylor: ***Yeah, and if we even out the towers until they are all the same height we can find the mean.***

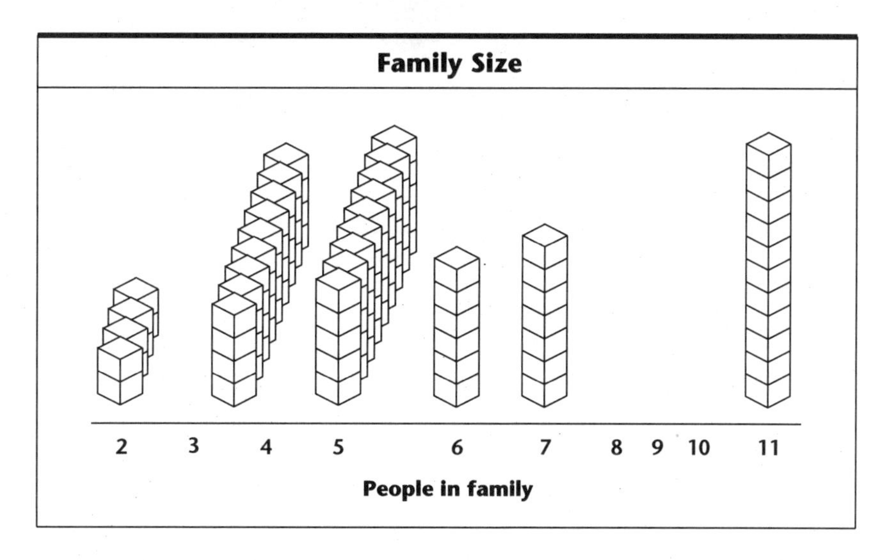

When the towers are made the same height, there will be 13 extra cubes. These cubes would be shared among the 25 towers, yielding a decimal.

Taylor added values on the horizontal axis, but did not consider how many times each value occurred in the data. Using cubes, he can see how this is done.

Vignette: Interpreting Scatter Plots

Overview

Sometimes it is useful to know the relationships between two variables, such as various body parts. For example, police try to determine how tall or how heavy someone is by measuring length, width, and depth of a footprint. Determining height or weight from footprints is possible because of *mathematical modeling.* Mathematical modeling fits a mathematical equation to data. Scatter plots are one way to represent such models. What do the points on the graph of such a model really mean? The following vignette illustrates several naive conceptions about interpreting points in a scatter plot.

Vignette

Women in Dr. Overby's college class measured the lengths of their left feet and their heights. They graphed their data on a scatter plot with height as the *x*-axis and length of left foot as the *y*-axis.

Dr. Overby's Students' Data

Height in Inches	Length of Left Foot in Inches
59.5	9.62
66.5	9.12
60	9
66.67	10.12
62.5	9
64.5	9.5
65.62	9.87
63	9.37
64	9.5

Most of the data points appeared to lie close to a straight line. Dr. Overby's class used computer software to produce a scatter plot and the *line of best fit*. Then Dr. Overby labeled a point *A*.

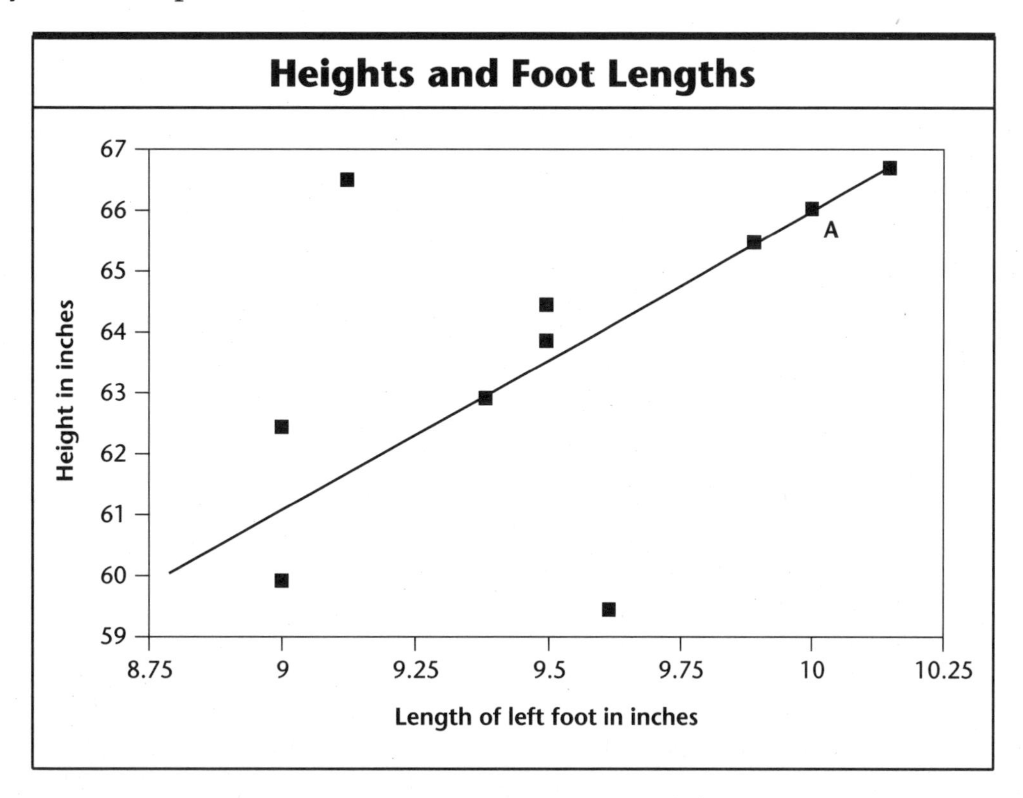

Dr. Overby:	***What does a point on the line mean in this investigation?***
Juanita:	***Point A stands for someone in our class who has a left foot of 10 inches and a height of 66 inches.***
Bonnie:	***Well, point A could just mean that someone has a left foot that is 10 inches long. I don't think it means any more than that.***
Dilek:	***I sort of agree with Bonnie, except that it means the person's left foot is between 10 and 66 inches long.***

Discussion Questions: Interpreting Scatter Plots

1. What are the students' different conceptions of point *A*?
2. What impact do these conceptions have on the students' interpretations of their class's data?
3. How can Dr. Overby best help each student?

Sample Responses to Discussion Questions: Interpreting Scatter Plots

1. **Juanita**

   It is important to realize that a point on the straight line in the data set may not be associated with an actual person. Rather, the model represents a relationship between the variables, so that a point might better be thought of as a *hypothetical* person. The model, or relationship, is useful for predicting the value of one variable when a particular value of the other variable is known, but these predicted values may not actually be found as an actual person's body measurements.

   Bonnie and Dilek

   Bonnie and Dilek both interpret points on a best fit line for one axis (that is one variable) and not for the other axis simultaneously. That is, Point A's x and y coordinates might mistakenly be interpreted as representing only a value of x or only a value of y. In other words, interpretations of the data might involve only individual variables and not the *relationship* between the variables. Bonnie and Dilek both use only one variable—the variable of foot length—to interpret the graph. They miss the fact that a scatter plot represents the relationship of two variables, in this case the relationship between foot length and body height. Although Dilek uses numbers from both axes of the graph, she thinks of those coordinates as values of only one variable—foot length.

2. If points on the line are interpreted as real people, there may be confusion about the nature of the data set used to generate the scatter plot. This confusion may interfere with making appropriate generalizations about the data.

3. Dr. Overby could ask students to select several points on the line of best fit, write down those x and y values, and then find out if people in the class really have those combinations of height and foot length. Students will discover that a point on the line does not represent an actual person.

Vignette: Unequal-Size Groups

Overview

Mr. West's students are designing an investigation around the question, "Who can jump rope more times without stopping—the teachers or the students in our class?" The children believe they have a problem, in that the class has fewer children than the number of teachers in the school.

Vignette

Cory: ***I'll bet we can jump* more *times than the teachers. Most of them don't even wear sneakers.***

David: ***Yeah, but there are a lot of teachers. They'll have more jumpers than we have.***

Cory: ***That's not fair.***

Miranda: ***That is certainly not fair. Maybe some of us can jump twice or some of them just won't jump at all.***

Cory: ***It's not fair to have some people jump twice.***

Miranda: ***We need to have the same number of people in both groups.***

David: ***How many people are in our class?***

Miranda: ***Twenty-two.***

David: ***Mr. West, how many teachers are there in the whole school?***

Mr. West: ***Twenty-seven.***

David: ***So subtract 22 from 27. They have five extra teachers.***

Miranda: ***How do we decide which five won't be allowed to jump?***

Cory: ***Maybe the first 22 who show up will get to jump.***

David: ***OK. Then we'll have 22 teachers and 22 of us.***

Discussion Questions: Unequal-Size Groups

1. What is the students' present conception about comparing groups?
2. What conception about comparing groups do the students need in order to allow *all* the teachers to jump rope?
3. What questions could Mr. West ask to help the students extend their conception about comparison of groups?
4. Should Mr. West intervene? Why or why not?

Sample Responses to Discussion Questions: Unequal-Size Groups

1. The students believe that only groups of equal size may be fairly compared.

2. The students need to learn that measures of central tendency—mean, mode, median—and range allow us to compare groups of unequal size.

3. Mr. West could ask questions about the kinds of conclusions the students expect to make.

   - If just one teacher jumps better than any of you in this class, will that mean that teachers, as a group, are better jumpers than students?

   - What data will it take to convince you that the teachers are the better jumpers?

   Mr. West may introduce the word *typical*, as in "Are you comparing one teacher and one student, or are you comparing the typical teacher and the typical student? How could we find a 'typical' teacher?"

4. Whether or not Mr. West intervenes at all may depend in part on how many new processes the students are already using in this investigation. Certainly the students have presented a teachable moment, and Mr. West has an opportunity to use the context of the students' questions to help them extend their concept of comparison between groups.

   - Why do you say that it is "not fair" to have more teachers than students?

   - Why do you need the two groups to be the same size?

Vignette: Line Graphs

Overview

Line graphs are used to show data and their change over a specified period of time. These are known as *time series graphs*. The x-axis shows time intervals, and the y-axis shows the variable that is being measured. Data are marked on the grid using points, and those points are connected with straight line segments. Drawing straight line segments between data points means you are assuming that there could be additional data points between any two connected data points. Drawing a straight line between two data points also implies that the change was steady between those two points.

Examples of data that may be represented with line graphs are

- height of a student over several years
- speed of a car over some distance
- temperature over several days

As we select ways to represent data, we reveal our assumptions about that data. Looking at graphs created in your workshops will help you to identify conceptions held by participants and to decide when and how to extend those conceptions. In this vignette, Anna is facilitating a workshop for teachers. The teachers draw a graph that reveals certain ways of thinking about their data.

Vignette

Anna asks teachers in her workshop to investigate sugar content in breakfast cereals. To find out if there is a relationship between sugar content and shelf location, Anna asks each group to represent cereal sugar content according to the location of the cereal on the top, middle, or bottom supermarket shelf.

In reporting their work, Mike, Sandra, and Barbara show the graph on the next page.

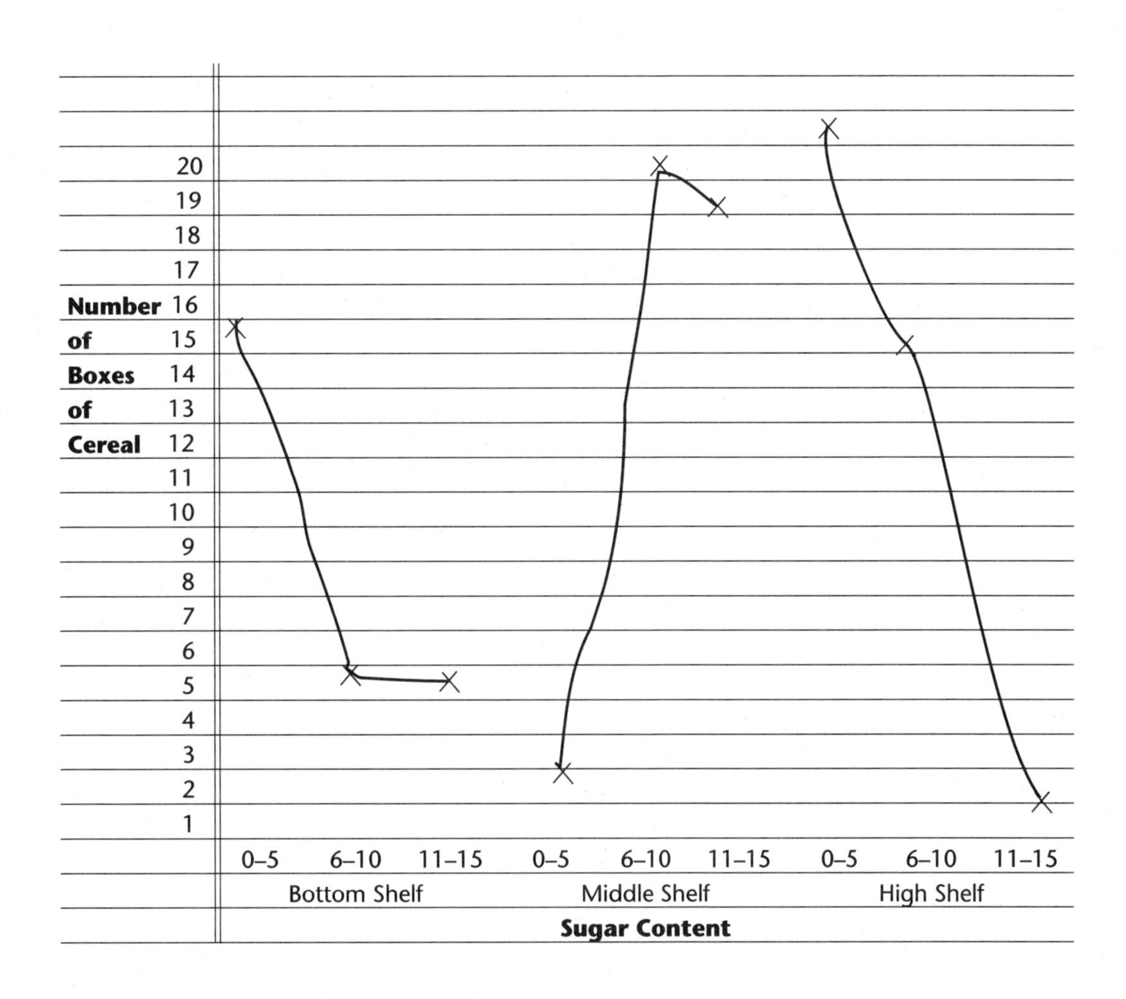

Discussion Questions: Line Graphs

1. By looking at their graph, what would you say Mike, Sandra, and Barbara seem to understand about using graphs to represent data?
2. What do you think they are ready to learn next about line graphs?
3. What teaching questions could Anna ask this group of teachers?

Sample Responses to Discussion Questions: Line Graphs

1. Mike, Sandra, and Barbara have found a way to group the data by using intervals of grams to refer to sugar content. By combining the data and displaying it across one *y*-axis, they make it easy to compare the three different shelves of cereals. However, they have inappropriately used lines to connect the points.

2. These teachers are ready to consider

   - that intervals are usually the same size. Note that 0–5 is an interval of six and 6–10 is an interval of five.

   - that they are graphing the *number of boxes* with a sugar content falling within a given interval of grams. They are not looking at data that are changing over time.

   - that because grams of sugar are continuous data, they can use a histogram to show the data for each shelf, using equal intervals such as 0–4, 5–9, 10–14, and 15–19 for each graph.

   - that three boxplots, one for each shelf's data, may be used to compare these data sets.

3. Anna might ask questions such as, "What other representations could you make for these data?" "How might you use histograms or boxplots?" She might provide a discussion of what a line graph is intended to show and then ask if these data "fit" this definition.

Vignette: Probability

Overview

The game Markers on a Line allows us to explore concepts of probability in a problem-solving atmosphere. Players are given a grid, numbered from 2 to 12, and 11 markers to distribute across the grid. Players may place as many markers as they want on each number—one, more than one, or none at all. Two dice are rolled, and their sum determines which markers may be removed. If a total of 8 is rolled, for example, one marker may be removed from the 8. The game continues until all the markers are removed from one player's grid.

In a Teach-Stat workshop, teachers may play this game several times, first in a large group, then in pairs. Based on their understanding of probability, teachers develop strategies for placing markers. Some teachers may not recall that the possible sums of two rolled dice are not all equally likely. These teachers may initially place one marker on each number. Other teachers, who know from experience that 2 and 12 are rarely rolled, may not place any markers on these two numbers and then may distribute the remaining markers evenly across the other numbers. Teachers who know the probability of rolling a 7 is greater than the probability of rolling any other number may place most of their markers on 7.

Vignette

Markers on a Line was one of the first probability activities presented at a Teach-Stat workshop. That evening, teachers wrote in their journals about any confusion they may still have had about the game. One of the teachers, Jerry, wrote the following:

Markers on a Line: Now I am confused! I studied probability in one of my college math courses three years ago. I liked it a lot and even made a B in the course. For the last two years in my class I have made sure that I included probability sometime during the year. To make a long story short, I'm pretty good at it, and I certainly know a lot about dice. I know that the probability of rolling a 7 is $^6/_{36}$, the probability of rolling a 6 or an 8 is $^5/_{36}$, and so on, until the probability of rolling a 2 or a 12 is $^1/_{36}$.

I started off by placing four of my markers on 7, two markers each on 6 and 8, and one each on 4, 5, and 9. I LOST! So then I placed three markers on 7, two on 6 and 8, and one each on 4, 5, 9, and 10. I LOST AGAIN! My partner always won, and she used weird strategies. Once she even placed one marker on each of the numbers from 2 to 12, and she still won! What's going on here?! I am going to have to think about this some more.

Discussion Questions: Probability

1. Why is Jerry confused about the results of playing several rounds of Markers on a Line?

2. What could the workshop leader do to help Jerry?

Sample Responses to Discussion Questions: Probability

1. Jerry's misconception is sometimes called *gambler's fallacy* and is common among adults. People believe that the results of just a few trials should be very close to the theoretical probability. It is important to stress to teachers and students that probability deals with phenomena that are random. In other words, results of the game exhibit a pattern over many repetitions, and patterns may not appear over a few trials.

2. Jerry may benefit from playing more rounds of the game or from pooling his game results with those of other teachers.

Notes: Misconceptions of Probability

1. *Gambler's fallacy* is the belief that results of a few trials should reflect the theoretical probability of an event.

   Example: A coin is tossed six times, and a head comes up on each toss. If a student believes that on the next toss a tail is more likely than a head, the student is operating under this misconception.

2. Some people believe that all possible outcomes of an experiment are equally likely.

   Example: Three coins are tossed. The misconception is that three tails are just as likely to occur as two heads and one tail. These two outcomes are not equally likely.

3. Another source of confusion about probability is insensitivity to the effects of sample size on predictive accuracy.

   Example: The chance that a baby will be a boy is about 50 percent. Over the course of an entire year, would there be more days when at least 60 percent of the babies born were boys in a large hospital, in a small hospital, or would the size of the hospital make no difference? Many students believe the hospital size would not make a difference. However, there is a good chance that a small hospital with only a few babies born each day would have more days with 60 percent boys.

4. A common pitfall is the tendency to make predictions or to estimate probabilities based only on personal experience or on the ease with which instances of an event can be constructed.

Example: If people are asked to estimate the divorce rate or the probability of being involved in a car accident, their estimates may be based on their own personal experiences with divorce or accidents.

References

Bright, G. W. "Measurement, Probability, Statistics, and Graphing." Chapter 5 in *Research Ideas for the Classroom: Middle Grades Mathematics*, ed. D. T. Owens. New York: Macmillan Publishing Company, 1993.

Shaughnessy, J. M. "Misconceptions of Probability: From Systematic Errors to Systematic Experiments and Decisions." Chapter 14 in *Teaching Statistics and Probability*, ed. A. P. Shulte. Reston, Virginia: National Council of Teachers of Mathematics, 1981.

Shaughnessy, J. M., and B. Bergman. "Thinking About Uncertainty: Probability and Statistics." Chapter 10 in *Research Ideas for the Classroom: High School Mathematics*, ed. P. S. Wilson. Reston, Virginia: National Council of Teachers of Mathematics, 1993.

Vignette: Responding to Participants' Questions

Overview

Teachers often report that they feel a bit anxious when preparing to teach a new topic or content area. "Suppose a student asks me a question for which I don't know the answer? What should I do or say?" These questions are valid, especially for teachers who are not confident about their background in mathematics.

The following vignette illustrates two points. First, teachers of young students need some understanding of statistical concepts beyond the grade level of their students. Second, having *the* mathematical "answer" is not always necessary when teaching statistical concepts.

Vignette

One day Cari, an elementary-school student, brought a two-liter soda bottle with dried beans in it to school.

Cari: ***Mrs. Chung, I have a probability question. In this bottle there are six different colors of beans, and I put in 11 of each color. Suppose I want to shake out four beans of the same color. How many shakes will it take us to get that?***

Mrs. Chung began to get a bit nervous. She still had a lot of questions herself about probability. Should she make a tree diagram? Was there some sort of formula for this? What *is* the "right answer"? What should she do?

Discussion Questions: Responding to Participants' Questions

1. What would you do if you were Mrs. Chung?
2. In general, how can you respond to students who ask questions that you feel are beyond your immediate content knowledge?
3. How might you respond to teachers at a workshop who ask questions about content that you feel is beyond your immediate knowledge?

Sample Responses to Discussion Questions: Responding to Participants' Questions

1. What Mrs. Chung actually did in this situation was effective. She asked her students to predict how many tries it would take to get four beans of one color. Then they conducted the investigation. They found it took 61 tries to get four beans of the same color.

   Was this result typical? We don't really know, because 61 could be an outlier. To find out, we could repeat the investigation many times and plot the outcomes on a line plot.

   Were the results biased? We don't know, but to make sure that each group of four beans was randomly selected, the beans shaken out would have to be replaced in the bottle and the bottle thoroughly mixed each time.

   This question can also be investigated by working a simpler problem. Instead of filling the bottle with six different colored beans, try it first with four beans or chips—two of two different colors. What is the probability in this case of getting two chips of the same color? (two out of four outcomes, or ½)

2. Possible responses to students could include the following:

That's an interesting question, Cari. Class, how could we investigate that?

I'm so glad you asked us that question, Cari. How would you propose we investigate this?

   The investigation could be followed up the next day after the teacher has had a chance to read about the content that night. The follow-up might include a review of the question, a summary of the investigation, and presentation of another way to approach the content or problem.

3. Possible responses to teachers at a workshop could include those below.

That's a good question, Wanda. I'm not really sure about that. Lynne, can you help me out here?

That's a good question, Wanda. Let me make a note here, and I'll find out for you tomorrow.

That's a good question, Wanda. Let's look in this statistics text and see if we can find what you're looking for.

MODULE 7

Presentation Skills

Overview

In *Activity 1, Introducing Constructive Criticism,* statistics educators discuss ways to analyze strengths and problems in workshop presentations. An optional activity for sparking discussion is a humorous demonstration of how *not* to present in workshops.

Activity 2, Rehearsing and Exchanging Constructive Criticism, presents a model for practicing introductions and conclusions to Teach-Stat investigations and for exchanging constructive criticism among statistics educators.

Activity 3, Daily Debriefing Sessions, prepares statistics educators to respond to feedback from colleagues and workshop participants. Strategies are discussed for helping unhappy workshop participants.

Goals

Statistics educators will

- improve their presentation skills
- improve their ability to give and receive constructive criticism
- learn more about the investigations for which they are understudies
- develop strategies for responding to troublesome workshop participants
- develop ways to respond to participants' daily feedback

Materials

- transparencies and content for the staff developer's example of a poor presentation (optional, *Activity 1*)
- *Don't Spit on the Overhead: Tips for Effective Presentations* (Handout 7.1)

Notes

Activity 1: 45 minutes

Activity 2: 60 minutes

Activity 3: 30 minutes

Materials

Statistics educators may wish to use video cameras in *Activity 2* when they rehearse their presentations.

Notes

Without appropriate training, teachers avoid confronting one another about mistakes and may resist exchanging constructive criticism. This aspect of school culture has been interpreted as an obstacle to implementation of lead-teacher programs in elementary schools in North Carolina. We don't know if this is a characteristic that can be generalized across the country. (Vesilind, E. M., and M. G. Jones. *Through a Sideways Door.* Chapel Hill, North Carolina: University of North Carolina Mathematics and Science Education Network, 1993.)

In one Teach-Stat workshop, a statistics educator stepped in and corrected an error her partner had made during a presentation. This occurred only after the staff developer indicated *he* would not intervene. The staff developer's decision not to intervene led to the statistics educator's increased confidence and appreciation of the roles of partners and constructive criticism.

In another workshop, statistics educators agreed ahead of time to help each other during the presentations. A presenter could say, *Help me out, group,* and another statistics educator would step in with the needed example, question, or explanation.

- video taping and playback equipment and tapes (optional for *Activity 2*)
- *Introducing and Concluding an Investigation* (Handout 7.2)
- *Unhappy Campers* (Handout 7.3)
- *Sample Strategies for Unhappy Campers* (Handout 7.4)
- one set of *Sample Comment Cards* (Handout 7.5)

Facilitating *Module 7*

Statistics educators need to know that you expect them to prepare carefully for their roles in the workshop and that you will provide relatively safe opportunities for them to rehearse and receive suggestions from each other. Express your belief that everyone, including you, can improve presentation skills. By modeling self-evaluation and openness to criticism, and by setting ground rules for feedback, you can help statistics educators feel more secure about exchanging feedback. If this exchange begins before the workshop, then daily debriefing sessions during the workshop will be more productive.

Activity 1 Introducing Constructive Criticism

One way to begin this activity is for you, the staff developer, to deliver a poor presentation to statistics educators. You may wish to use the first five to ten minutes of a Teach-Stat investigation from *Teach-Stat for Teachers: Professional Development Manual.* For example, you might fail to plug in the projector, mix up the transparencies, stare at the ceiling, speak too fast, and so on. Do not tell statistics educators that you are trying to give a poor presentation. They will eventually laugh and grasp your intent!

Introduce the activity by saying something like this

One way we can develop presentation skills is to receive feedback from friends who watch us practice. Today I am going to show you a way of exchanging constructive feedback. Then you will have an opportunity to try this as you rehearse for the workshop. During the workshop we will meet at the end of each day. One purpose of those meetings will be to exchange feedback that will help the following days be even better. So, it's important that we think now about giving and receiving feedback. To begin, I am going to give a brief presentation to you, and then I will ask you to give me constructive feedback.

After your presentation to statistics educators, model the following way to exchange feedback.

The model of exchanging feedback that I am recommending is this. First you will tell me at least one positive thing about my presentation. Then I will ask you a question to elicit a suggestion for improvement. You will not give me suggestions unless I ask for them. Let's try this. You begin now by telling me one positive thing about my presentation.

Begin with positive feedback. Encourage specific and meaningful feedback. Tell the group which comments were most helpful to you as a presenter. Comments such as "You were great!" are not as helpful as more specific comments, such as "Your example for grouped data was clear."

When you are satisfied that statistics educators have given specific positive feedback, then go to the next step—asking for suggestions.

First, I'll ask you how my pace was. Did I go too slow or too fast?

Encourage statistics educators to offer several different answers, so that the group can discuss the most helpful, friendly style of feedback. Again, insist on useful, specific feedback.

How engaged did you feel with what I was saying? Why? Which did I seem to do more often—tell answers or ask questions?

My last question for feedback is this: what can I do to improve my presentation?

Point out that this model of feedback allows the person receiving feedback to be in control of the topic. The presenter will receive only as much feedback as she or he is ready to ask for. The last question asked shows readiness to hear feedback on any topic.

Helpful feedback can also be given when statistics educators who are not presenting describe what the workshop participants were doing during the presentation. An example of such feedback is, "Marcia, when you moved on from the line plot to the graph, two people back here still had questions, and I'm not sure you saw their hands." This gives Marcia a chance to decide whether she should meet individually with those two participants. The feedback also lets Marcia and other statistics educators know that pausing for questions and watching for raised hands are important.

Summarize *Activity 1* by discussing *Don't Spit on the Overhead: Tips for Effective Presentations* (Handout 7.1). Encourage statistics educators to add items to the list.

Notes

Note how the presenter prompts the continuation of feedback by asking specific questions. This gives the presenter some control over the amount and type of feedback he or she is ready to receive.

Statistics educators report that promoting discourse among workshop participants, as opposed to telling or lecturing, is a difficult technique to master. As statistics educators rehearse and observe each other, help them to practice questioning strategies.

Notes

Some statistics educators prefer to rehearse alone before they feel ready to present to someone else for feedback. It is important for each statistics educator to discover what methods of rehearsal are most effective for her or him.

In one Teach-Stat workshop, dialogue journals helped statistics educators to evaluate participants' daily progress and plan for the next day. Participants wrote each day about one focus question (for example, *How would you teach the concept of mean in your grade level?*) and one open question (for example, *How are you doing today?*).

Each statistics educator was responsible for reading four teachers' journals. Statistics educators wrote their responses to the teachers' journals on stick-on notes so that the staff developer could suggest possible revisions of the responses.

Activity 2 Rehearsing and Exchanging Constructive Criticism

In this activity, statistics educators practice presentation skills by working in pairs made up of investigation leaders and understudies. Statistics educators responsible for each investigation on the workshop agenda now prepare in detail the way they will actually introduce and conclude their investigations.

The handout *Introducing and Concluding an Investigation* is designed to guide this process (Handout 7.2). Introductions and conclusions may be rehearsed, shared, and videotaped, so that statistics educators will have opportunities to reflect on and to revise their presentations before the actual workshop.

The understudy may either observe or videotape the leader. After each microteaching, the pair exchanges feedback in the way modeled in *Activity 1*. In addition to providing practice for leaders, this activity helps each understudy to grasp the overall plan for that investigation in case they need to substitute for the leader.

Activity 3 Daily Debriefing Sessions

A powerful professional development experience for statistics educators has been the daily debriefing sessions held at the conclusion of each day of the workshop. Purposes of debriefing sessions include

- evaluating feedback on participants' comment cards
- giving and receiving feedback with each other about presentation skills
- reviewing plans for the next day

As staff developer, your role in the daily debriefing sessions is to provide an objective overview of the workshop in progress. When comment cards are used, watch for the tendency for statistics educators to focus their reflections only on the comment cards.

Sometimes the feedback can be harsh; sometimes not harsh enough. One way to edit comment cards is for you to read the cards privately while statistics educators clean up the room or prepare for the next day. If any comments seem too harsh, too personal, or unfounded, you can put them aside. You may wish to discuss some comments on an individual basis with each statistics educator.

Another topic to address during debriefing sessions is unhappy workshop participants. A little preparation can go a long way toward helping statistics educators cope effectively with workshop participants who appear withdrawn, disgruntled, or even hostile. Statistics educators need to know that workshops occasionally contain an unhappy camper, that workshop leaders need not view the behavior as personal attacks, and that experienced workshop leaders have a repertoire of strategies for responding to such participants. By discussing vignettes and comment cards in the course of this activity, statistics educators can share ideas for confronting behavior that threatens the success of the workshop. In addition, several suggestions are given below.

Sometimes we find "unhappy campers" in our workshops. I'm telling you this now, so that you won't take it personally if we do have such a teacher in our workshop. There are several styles of unhappy campers. Some people will withdraw from the group or try to get others to join them in a withdrawn little group in the back of the room. Others may sit up front and disagree with much of what is presented. Sometimes we learn that these teachers were required to attend or have signed up just for the money. Some teachers may chatter through presentations or dominate discussions. Whatever the situation, it's important to deal with it immediately, so that the tone of the workshop remains enthusiastic, friendly, and safe for intellectual risk taking.

In general, statistics educators should confront troublesome participants, rather than simply complaining about them to each other.

Lead statistics educators through a reading and discussion of *Unhappy Campers* (Handout 7.3) and/or *Sample Comment Cards* (Handout 7.5), depending on the time available. Strategies for confronting unhappy workshop participants are suggested in *Sample Strategies for Unhappy Campers* (Handout 7.4). Statistics educators may be able to generate and share other strategies.

Notes

At one Teach-Stat workshop, statistics educators placed participants' name tags at tables as a way of dividing little groups. By also placing their own name tags at the tables, statistics educators were able to monitor participants effectively.

Don't Spit on the Overhead: Tips for Effective Presentations

Here are tips for effective presentations. From your own experiences, you may wish to add to this list.

Exude confidence and enthusiasm. Create a positive impression from the moment you enter the room. All of what you do and say, even at lunch and in restrooms, affects the level of confidence participants will have in you. An off-hand comment about stage fright or misplaced materials might affect the success of a workshop.

Have materials ready. Transparencies should be in sequence, and equipment and software tested ahead of time. A high level of preparedness allows you to interact more with participants and lets them feel more at ease.

Practice with the overhead projector. Make sure all data sets, line plots, or graphs you will draw during your presentation will fit legibly on a transparency.

Be gentle and approachable. A smile signals many positive messages, including a willingness to get to know participants and their issues.

Interact with all participants, not only those you already know or feel most comfortable with. Try to draw all participants into discussion and activities.

Practice pronunciation. If unsure about the pronunciation of a word, look it up before the workshop. Be sure you are pronouncing participants' names correctly, too.

Speak at a moderate pace and with a modulated, not monotone, voice.

Maintain eye contact with participants. This holds attention and helps you to notice any confused or questioning looks the participants may be giving you.

Think out loud when you model. Remember that you are modeling a thinking process and not simply an act of building block towers or drawing graphs.

Pause at appropriate times for questions.

Affirm relevant questions with, "That's a good question."

Plan and carefully practice directions you will need to give. This is especially important when launching investigations or other extended activities. Unclear directions can result in false starts and wasted time. Don't assume that you will remember how to present something you have previously presented.

Plan conclusions of investigations and other activities carefully. There should be a summarizing of key points and insights. Relate the investigation to other parts of the workshop and to goals and objectives. As a clue for your audience, begin this part with a phrase such as, "To summarize, ..."

Praise participants when they achieve objectives and when they struggle constructively with tasks. Participants like to know how they are doing.

Maintain an optimistic attitude when confusion arises about a statistics concept. If you are not sure how to allay the confusion, tell teachers that you will prepare a different example for the next session and that you will return to the confusing topic. It is fine to say, "This is an interesting question, and I need more time to think about it."

Divert an overly talkative participant by saying, "I'd like to continue this topic after this session. Can we talk then?"

Move around the room, when possible.

Introducing and Concluding an Investigation

On index cards, write out your introduction, including directions you will give to get workshop participants started on one of the investigations you are facilitating. Read or tell your introduction to your understudy or partner. Ask your partner to suggest improvements. Even if you do not read the cards during the workshop, you will be more confident and give clearer directions having written them out ahead of time.

Use *Teach-Stat for Teachers: Professional Development Manual* as a source of ideas for facilitating your investigation. Effective introductions and conclusions to investigations may include

- pointing out how the investigation is in line with local curriculum guidelines
- pointing out how the investigation could help students on end-of-course tests
- sharing of students' work from your own classroom, with discussion about levels of conceptual understanding revealed in the students' work
- asking teachers to share ideas for interdisciplinary uses of the investigation

Although you may have successfully taught the investigation to your own students, you will need to practice explaining key concepts to an adult audience. Decide exactly which parts of the process of investigation you will emphasize. Practice with an overhead to determine how much of a data set or graph you can clearly put on one transparency.

If possible, videotape your dress rehearsal of the introduction to the investigation. A tape will help not only you but could also be used by your partner or understudy.

Unhappy Campers

As a statistics educator facilitating a workshop, how might you respond to the participants described below?

Helena sits in the front of the room. She has brought her own graphing calculator and works problems on it during the presentations. Now and then she raises her hand to disagree with the presenter. Most of her comments begin with "That isn't what I get" and "A better way to do it is . . . ."

Wilma sits at the back table. She has never contributed or asked a question. She does not write in her notebook or participate much in her small group. When the group members ask her to help, she says, "Whatever you want to do is fine with me." She frequently looks at her watch. At the end of each session she is first to leave, practically running out the door.

Sandy is one of the few men in the workshop. He is good looking, amusing, and popular among other participants. At his table there is a great deal of chatter, even while presenters are giving directions and minilectures. During the morning break and lunch, Sandy is the star joke teller of a small group that seems to try to be together all the time. In the afternoon at Sandy's table there is giggling and note passing. Several participants at other tables begin to watch Sandy to find out what is so funny.

Sample Strategies for Unhappy Campers

The Heckler: Meet this person at the first break. In a calm voice say, "I get the feeling this workshop may not be for you. What do you think?" This lets the person know that you have noticed the heckling behavior and will not passively tolerate it.

The Withdrawn: During a break or at lunch say, "Something seems to be bothering you."

The Chatterer: All statistics educators, especially those not presenting at the time, can help with chatterers. Stand near them. Sit down with them. Whisper to them, "I can't hear the presenter. Can you keep it down? Thanks."

Sample Comment Cards

The room is too cold.

I went to a graphing calculator workshop and learned things that don't fit with what you're saying here. I feel confused.

We like working in our group.

Could you please have caffeine-free diet soda and not just diet soda?!?

When do you use box and whiskers? Am I the only one who didn't "get" this??

Our group is great!

Our group is not so great. One person has strong opinions, and we do everything her way.

Thanks for everything you're doing—great job!!!!

I came to this workshop to learn more about bivariate data, but I'm worried that we aren't going to get to that topic.

You're moving too fast for me. Today I was lost from the beginning.

I've already used most of this stuff, but you're doing a good job teaching the others.

I hate writing in the journal. I just can't tell you what I'm thinking this way.

MODULE 8

Preparing the Site

Overview

During *Activity 1, Orientation to the Site and Support Staff*, statistics educators gain familiarity with the workshop site and learn protocol for using photocopy machines, telephones, supply rooms, and other resources at the site.

In *Activity 2, Organizing the Room*, statistics educators plan the most effective use of the space available.

Activity 3, Final Preparations, is a time for statistics educators to set up equipment and materials (if a site visit is made).

Module 8 concludes with *Activity 4, Inspiration*, a brief sharing of encouragement and excitement about the upcoming workshop.

Goals

Statistics educators will

- prepare a physical setting for a Teach-Stat workshop
- effectively interact with support personnel at the workshop site

Materials

- workshop equipment and materials, as ordered by statistics educators
- photocopy machine
- access to and cooperation from office staff and other support people, such as custodians
- staff developer's own list of site-specific preparation tasks
- floor plan and photographs of the site (if a site visit is impossible)

Notes

Activity 1: 15–30 minutes

Activity 2: 15–30 minutes

Activity 3: 60 minutes

Activity 4: 10 minutes

It is helpful, but not necessary, to present *Module 8* the day before the workshop and at the workshop site.

Familiarity with the site helps statistics educators to visualize how the investigations will happen in the given space and contributes to their confidence.

If a site visit is impossible, statistics educators can work through *Module 8* with a floor plan and photographs of the site.

Facilitating *Module 8*

In preparing the site, statistics educators in a new way assume responsibility for the workshop. They may be both excited and apprehensive. Time should be provided for statistics educators to walk through their roles in the space and with the equipment. This is especially important for those who will facilitate activities on the first day of the workshop. As the workshop continues, routine uses of space, furniture, and equipment will become established.

Activity 1 Orientation to the Site and Support Staff

A workshop is often conducted at a site that is unfamiliar to the workshop presenters. As a staff developer, you should arrange to introduce statistics educators to administrative and support staff at the site. Determine how statistics educators may call on this staff for assistance.

Discuss access to telephones, both for statistics educators and for workshop participants. Decide who will have access to photocopy machines and paper.

Orient statistics educators to the building, pointing out supply rooms, audiovisual equipment storage, computer lab, smoking areas, and restrooms. Discuss the security of equipment and materials.

Notes

Check to see if signs to the workshop are in place.

On the first day of a workshop, participants often arrive very early, especially if they are unfamiliar with the site. On the first day all presenters should be at the site one hour prior to the starting time of the workshop.

Activity 2 Organizing the Room

With the statistics educators, move into the space where the workshop will be conducted and discuss the following items:

Where will we stand to present?
Where will the teachers sit?
Where will small groups work together?
Which walls or bulletin boards will we use?
Do we need chalk or markers?
Is there a projection screen?
Where is a good place for the snack table?
Are electrical outlets where we need them, including an outlet for the coffee pot? How many extension cords will we need?
Can we set up a supplies table for graph paper, stickers, and all the other raw materials for investigations?
Where will teachers go during their breaks?
Will there be a registration table?
Where will teachers get their name tags when they arrive?
Will we have packets on the tables before teachers arrive?
Do we have trash cans?

Is there a thermostat? Do windows open?
Where are emergency exits?
Where will we statistics educators sit when we are not presenting?

Activity 3 Final Preparations

After the physical environment has been discussed and arranged, statistics educators may need time to take care of individual tasks and assignments. Tasks may range from physically setting out materials, to once more reviewing a certain statistical concept, to making copies of handouts and putting together participants' packets.

Activity 4 Inspiration

As presenters prepare for opening day of the workshop, there may be a tendency to become overly focused on details and to lose sight of the big picture. Close this preparatory session in an enthusiastic manner. One possibility is to read excerpts from *Teacher Talk* (Handout 1.4), such as Darcy's description of her growth as a teacher. You might thank statistics educators for all of their efforts up to this point and tell them how excited you are about the next part of your work together.

MODULE 9

After the Workshop

Overview

Module 9 is held after statistics educators have facilitated a Teach-Stat workshop for other teachers. *Activity 1, Evaluating the Teach-Stat Workshop and the Statistics Educators Institute,* is an opportunity to reflect on what went well during the workshop and what statistics educators would do differently next time. You may also ask statistics educators to evaluate the Statistics Educators Institute.

In *Activity 2, Planning Follow Up with Workshop Leaders and Participants,* statistics educators decide how to continue sharing and supporting each other after the formal Teach-Stat program.

Activity 3, Promoting Ourselves as Statistics Educators, helps the group look ahead to serving as workshop facilitators in schools.

Activity 4, Attending to Hidden Details of a Workshop, is a time to share information about tasks that may have been carried out behind the scenes by the staff developer but that statistics educators would need to do for themselves in the future.

Activity 5, Closure and Celebration, is intended to celebrate the network of statistics educators developed by this institute.

Goals

Statistics educators will

- evaluate the Teach-Stat workshop in which they taught
- evaluate their own experiences in becoming statistics educators
- discuss plans to follow up with workshop participants
- share ideas for promoting themselves as statistics educators
- learn from the staff developer about administration of the workshop, especially the budgeting process
- celebrate the culmination of a successful professional development endeavor

Notes

Activity 1: 30 minutes

Activity 2: 30 minutes

Activity 3: 30 minutes

Activity 4: 45 minutes

Activity 5: 15–60 minutes

Materials

- the staff developer's summary of workshop participants' evaluations
- the workshop budget (optional)
- *Evaluation Form: Statistics Educators Institute* (Handout 9.1)

Facilitating *Module 9*

Module 9 varies from site to site, as Teach-Stat staff developers try to meet needs and interests of each group of statistics educators. Activities described here have been used successfully at various sites.

Activity 1 Evaluating the Teach-Stat Workshop and the Statistics Educators Institute

Share with statistics educators the teachers' evaluations of the workshop. This may be done by reading aloud excerpts you have carefully selected to represent all the evaluations. Omit references to specific statistics educators. If you have evaluations about specific statistics educators that you think need to be shared, do this privately at a different time.

Include your own evaluations of the workshop, noting areas of professional growth and risk taking among statistics educators.

Ask statistics educators to reflect on their own professional growth and change during the Statistics Educators Institute. Use *Evaluation Form: Statistics Educators Institute* (Handout 9.1) to collect a formal evaluation.

Activity 2 Planning Follow Up with Workshop Leaders and Participants

Teach-Stat participants and leaders often indicate a desire to continue the professional development that has begun during their involvement in this workshop. If an interest is expressed, how and when will you plan for follow-up meetings? Does the budget provide for stipends for after-school or Saturday meetings? Will you plan dinners to share ideas? Are school visits, telephone conferences, or newsletters possible? Will you meet at a professional conference? All of these questions reflect follow-up strategies that have been successful at Teach-Stat sites.

Activity 3 Promoting Ourselves as Statistics Educators

Teachers who have successfully planned and implemented a Teach-Stat workshop are ready to continue their development as workshop consultants. Discuss with statistics educators what they individually or as a group would like to do to let principals and superintendents know that they are prepared to facilitate Teach-Stat workshops. Publicity suggestions include articles in state mathematics association newsletters, brochures handed out at conferences, and direct mailings to principals and district supervisors.

Statistics educators may also wish to discuss customary consulting fees in your locality and ways of negotiating consulting fees.

Activity 4 Attending to Hidden Details of a Workshop

As statistics educators begin to think about conducting Teach-Stat workshops on their own, they will have many questions. For example, how will they respond to a principal who asks about the cost of a Teach-Stat workshop? You might share information about the budget for the workshop in which statistics educators taught, and you can describe other administrative tasks that you did behind the scenes, such as applying for continuing education credits for the teachers, ordering materials, and finding a site. These are tasks that statistics educators may not have been responsible for this time but will need to know about in the future.

Notes

While discussing budgets, be sure statistics educators understand when and how they will receive any stipends and reimbursements from the workshop.

Other helpful issues to raise now include desired lengths of workshops. Rough outlines of one-hour, three-hour, six-hour, and one-week-long workshops could be discussed and developed as a group.

Handouts 2.6 and 2.7 may be discussed as sample agendas of three-hour and six-hour workshops.

Activity 5 Closure and Celebration

How you choose to celebrate the professional and personal growth of statistics educators depends on the group. Some groups celebrate with a potluck lunch or dinner; others exchange silly gifts. If you took photographs during the workshop, you might give photos to statistics educators or share a photo collage. Whatever the style of your group, be sure they close your time together with a sense of accomplishment.

Many groups of statistics educators make plans to meet informally over the next few months.

Evaluation Form: Statistics Educators Institute

Take a few minutes to reflect on your experiences in this program, and respond to the items below. Your comments will be used in planning future professional development projects.

1. Which experiences or activities in the Statistics Educators Institute were particularly helpful to you? Please explain your response.

2. Which experiences or activities in this program were least helpful to you? Please explain.

3. What question or concern do you still have about statistics and probability, or about facilitating a workshop about statistics and probability?

4. Other comments: